Membrane Nanodomains

Colloquium
Digital Library of Life Sciences

This e-book is a copyrighted work in the Colloquium Digital Library—an innovative collection of time saving references and tools for researchers and students who want to quickly get up to speed in a new area or fundamental biomedical/life sciences topic. Each PDF e-book in the collection is an in-depth overview of a fast-moving or fundamental area of research, authored by a prominent contributor to the field. We call these e-books *Lectures* because they are intended for a broad, diverse audience of life scientists, in the spirit of a plenary lecture delivered by a keynote speaker or visiting professor. Individual e-books are published as contributions to a particular thematic **series**, each covering a different subject area and managed by its own prestigious editor, who oversees topic and author selection as well as scientific review. Readers are invited to see highlights of fields other than their own, keep up with advances in various disciplines, and refresh their understanding of core concepts in cell & molecular biology.

For the full list of published and forthcoming Lectures, please visit the Colloquium homepage: www.morganclaypool.com/page/lifesci

Access to Colloquium Digital Library is available by institutional license. Please e-mail info@morganclaypool.com for more information.

Morgan & Claypool Life Sciences is a signatory to the STM Permission Guidelines. All figures used with permission.

Colloquium Series on Building Blocks of the Cell: Cell Structure and Function

Editor

Ivan Robert Nabi, *Professor, University of British Columbia,*
Department of Cellular and Physiological Sciences

This Series is a comprehensive, in-depth review of the key elements of cell biology including 14 different categories, such as Organelles, Signaling, and Adhesion. All important elements and interactions of the cell will be covered, giving the reader a comprehensive, accessible, authoritative overview of cell biology. All authors are internationally renowned experts in their area.

Published titles

(for future titles please see the website, www.morganclaypool.com/page/lifesci)

Membrane Nanodomains
John R. Silvius
www.morganclaypool.com

ISBN: 9781615046201 paperback

ISBN: 9781615046218 ebook

DOI: 10.4199/C00076ED1V01Y201303BBC001

A Publication in the

COLLOQUIUM SERIES ON THE BUILDING BLOCKS OF THE CELL

Lecture #1

Series Editor: Ivan Robert Nabi, University of British Columbia, Department of Cellular and Physiological Sciences

Series ISSN Pending

Membrane Nanodomains

John R. Silvius
Department of Biochemistry
McGill University

COLLOQUIUM SERIES ON THE BUILDING BLOCKS OF THE CELL #1

 MORGAN&CLAYPOOL LIFE SCIENCES

ABSTRACT

Many membranes in eukaryotic cells are inhomogeneous structures in which various membrane components are nonrandomly distributed, forming diverse types of 'domains.' Some membrane domains have long been well known, because they are sufficiently large, long-lived, and morphologically well defined to be characterized using classical microscopic and biochemical approaches. However, new technologies have revealed the presence in membranes of smaller, often highly dynamic 'nanodomains' that also play key roles in membrane function. Our current understanding of the diversity, the properties, and the functions of nanodomains is still very limited and, in some cases, controversial. Nonetheless, it is clear that many important aspects of membrane biology arise from features of membrane organization that 'play out' on spatial and temporal scales that are only now becoming experimentally accessible in living systems. In this book, we will discuss properties and interactions of membrane molecules that lead to nanodomain formation, new and emerging technologies by which nanodomains can be studied, and experimental examples that illustrate both highlights and current limitations of our present knowledge of the properties of membrane nanodomains in various cell types.

KEYWORDS

biological membrane, membrane lipids, membrane proteins, membrane curvature, lipid phases, lipid rafts, cholesterol, protein-protein interactions, protein-lipid interactions, fluorescence microscopy, electron microscopy, single-particle fluorescence, super-resolution microscopy, mass spectrometry, epithelial cells, lymphocytes, mast cells, cell signaling, membrane traffic

Contents

1. **Introduction** 1
    1.1   Notes on Terminology 2

2. **Thermodynamic Bases for Formation of Membrane Domains** 3
    2.1   Lipid Segregation in Model Membranes 3
    2.2   Lipid–Protein Interactions 8
          2.2.1   Interactions of Intramembrane Protein Domains with Lipids 8
          2.2.2   Interactions of Extramembranous Proteins and Protein Domains
                   with Lipids 11
          2.2.3   Determinants of Protein Association with Liquid-Ordered vs.
                   Liquid-Disordered Lipids 11
          2.2.4   Association of Proteins and Lipids with Highly Curved
                   Membrane Structures 16
    2.3   Protein-Protein Interactions in Membranes 19
    2.4   Membrane 'Scaffold'-Forming Proteins 22

3. **Kinetic Bases for Formation of Membrane Domains** 27

4. **Historical Development of the Concept of Membrane Nanodomains** 31

5. **Experimental Tools to Study Nanodomain Formation in Cells** 35
    5.1   Biochemical Fractionation Methods 35
    5.2   Fluorescence Microscopy and Colocalization Analysis 37
    5.3   Manipulations of Membrane Lipid Composition 39
    5.4   Lipid Model Systems 41
    5.5   Electron-Microscopic Methods and Analysis of Molecular Clustering 43
    5.6   Fluorescence Resonance Energy Transfer 49
    5.7   Bimolecular Fragment Complementation (BiFC) Microscopy 52
    5.8   Single-Particle Measurements of Molecular Diffusion and Interactions 52

5.9   Super-Resolution Light Microscopies... 57

5.10 Plasma Membrane-Derived Giant Vesicles... 61

5.11 Lipid Mass Spectrometry (MS).. 62

6. **Experimental Studies of Nanodomains in Cellular Systems**................ 63

6.1   Membrane Domain Organization in Fibroblasts 63

6.2   Nanocluster-Based Signaling by GPI- and Ras Proteins 65

6.3   Endocytic Trafficking of 'Raft' Components in Fibroblasts 69

6.4   Polarized Trafficking of Apically Localized Proteins in
      Renal Epithelial Cells.. 71

6.5   Activation of the T-Cell Receptor 73

6.6   Mast Cells and the FcεRI Receptor 81

7. **Looking Forward** 85

References 87

CHAPTER 1

Introduction

Many membranes in eukaryotic cells are not uniform, homogeneous structures but rather incorporate subcompartments ('domains') whose compositions and functional properties differ importantly from those of other regions of the same membrane. Some membrane domains are very large (with dimensions of microns or tens of microns), such as the apical and basolateral domains of the plasma membrane in epithelial cells, or the somatodendritic and axonal plasma membrane domains in neurons. Various types of smaller-scale membrane domains, such as caveolae, coated pits or ER exit sites, have also been clearly identified and extensively studied (see the books by Stan and Lamaze in this series). Characterization of these domains has been facilitated by their relatively large dimensions ($\geq$ca. 100 nm) and long lifetimes (seconds or longer), and by highly distinctive morphologies and/or protein compositions.

A major current research focus, and challenge, in membrane biology is to understand the nature and the functional importance of additional types of membrane domains that are smaller, and often more short-lived or spatially diffuse, than the well-established types of domains noted above. We will refer to these smaller structures as membrane 'nanodomains,' and they constitute the primary focus of this book. Both theoretical arguments and compelling, though limited, experimental data suggest that nanodomains contribute to membrane function in ways no less important than do larger, better-characterized domains like those noted above. At present, however, for both technical and sometimes conceptual reasons, characterizing the origins, the diversity, and the biological functions of membrane nanodomains remains one of the true frontier areas of membrane biology.

Frontier science is exhilarating but messy and often controversial; this is certainly true for the area of membrane nanodomains. Accordingly, this book will not comprise a set of 'textbook' examples illustrating clear, universally accepted roles for nanodomains in various aspects of membrane biology. Instead, we will here first consider some physical principles that can contribute to the formation of membrane nanodomains, then examine a number of experimental approaches used to investigate these domains and finally discuss some representative experimental systems that have made important contributions to our current (limited) understanding of them. Reflecting both the history of the field and much current interest, we will devote particular attention to one type of conceptual model for membrane nanodomains, the 'membrane raft' model. This model proposes that various membranes in eukaryotic cells incorporate sterol-dependent nanodomains, commonly termed 'rafts,' whose lipid components exhibit a physical organization distinct from that found elsewhere in the membrane. However, as we will see, there are many possible bases for formation of nanodomains within membranes, resting on a variety of types of interactions among proteins and lipids, and these can combined in many ways to create nanodomains whose diversity we may be only beginning to understand.

1.1   NOTES ON TERMINOLOGY

Nomenclature to describe membrane domains is sometimes poorly systematized, but we will use the following definitions throughout this book. 'Domains' will be used as a generic term to designate the full range of spatially differentiated regions within membranes, regardless of size or morphology. Domains larger or smaller than 100 nm will be referred to as 'microdomains' and 'nanodomains,' respectively. Finally, following current usage, we will use the term 'mesoscale' to denote the range of spatial dimensions from 10 nm (just above the dimensions of individual biomolecules) to roughly 250 nm, the limit of conventional optical resolution.

The widely used term 'membrane raft' has been interpreted in different ways as research in the field has progressed, and it remains difficult to define even today. Except where otherwise indicated, we will use the term here to indicate any form of structure, transient or stable, whose organization is based at least in part on a tendency of particular membrane lipids to self-associate preferentially in a manner that depends on cholesterol or related membrane sterols. It is important to note that the classifications of different membrane components as 'raft' or 'non-raft' species, or of a given membrane process as 'raft-dependent,' are often based on operational criteria of varying reliability. Caveolae have been suggested to constitute a specialized type of membrane raft. However, their compositions and functions are quite different from those of non-caveolar rafts, and we will focus on the latter in this book (see the book by Stan in this series for a discussion of caveolae and their properties).

• • • • •

CHAPTER 2

Thermodynamic Bases for Formation of Membrane Domains

Membrane 'domains' can in principle arise by two general types of mechanisms. The first is thermodynamic: particular subsets of membrane-associated molecules may tend to associate preferentially to form either small 'clusters' or larger segregated regions with distinct compositions within the bilayer. There is ample precedent for such behavior, as discussed in the subsections below.

2.1 LIPID SEGREGATION IN MODEL MEMBRANES

Biological membranes contain an enormous number of lipid species with different structures and physical properties [1]. Given this complexity, it was suggested as early as the 1970s that in some membranes lipids might form segregated but coexisting domains with distinct compositions and physical properties [2]. Consistent with this suggestion, experimental studies have demonstrated that lateral phase separations can occur in model membranes containing mixtures of lipids resembling those found in biological membranes. As discussed below, the separation of liquid-ordered from liquid-disordered phases has attracted particular attention for its possible relevance to 'raft' domain formation in eukaryotic cell membranes.

Bilayers prepared from lipids can exist in three different general types of phases [3, 4]. In solid (or gel) phases the lipid hydrocarbon chains are highly extended and closely packed together in an ordered, quasi-crystalline array, with very little freedom to 'flex' their hydrocarbon chains (Figure 1). Lateral diffusion of lipid molecules in solid-phase bilayers is extremely slow, and even rotation of lipid molecules about the bilayer–perpendicular axis occurs relatively slowly. By contrast, in liquid-disordered (Ld) bilayers the lipid hydrocarbon chains are more loosely packed and exhibit larger-amplitude flexing motions, and the lipid molecules rotate and diffuse laterally in the bilayer plane much more rapidly. In isolation, phospho- and sphingolipids with long saturated hydrocarbon chains can pack well into solid phases even at temperatures close to physiological and are sometimes referred to as 'high-melting' species. By contrast, phospholipids with *cis*-double bonds in one or both chains can assemble into solid phases only at temperatures well below physiological (for many such species, <0°C) and are termed 'low-melting' species. While liquid-disordered lipid phases are common in biological membranes, organisms typically tailor their membrane lipid compositions to minimize formation of solid lipid phases under physiological conditions. For this reason, segregation of solid- from liquid-phase lipids is not thought to contribute to membrane domain formation in most biological systems.

A third type of lipid bilayer phase, the liquid-ordered (Lo) phase, was first predicted by theoretical models and later demonstrated experimentally in lipid model membranes combining saturated phospho- or sphingolipids with cholesterol [4]. In the Lo phase, the lipid molecules diffuse laterally, and rotate about the bilayer–perpendicular axis, at rates only a few-fold slower than in the liquid-disordered phase. However, in this phase, 'flexing' motions of the lipid hydrocarbon chains are considerably more restricted, and the lipid hydrocarbon chains are on average substantially more extended than in the liquid-disordered phase (Figure 1). The distinctive properties of the Lo phase reflect in part the unique structural properties of cholesterol (and some related sterols, such as ergosterol). The multiple fused hydrocarbon rings of these sterols are structurally more rigid than the hydrocarbon chains of other membrane lipids; moreover, one 'face' of the fused-ring system

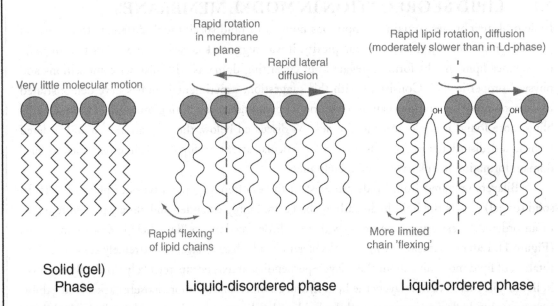

FIGURE 1: Lipid bilayers can exist in three types of phases. Solid or 'gel' phases exhibit tight, highly ordered packing of the lipid chains and very little molecular motion. Liquid-disordered (Ld) phases exhibit fast rotation and diffusion of lipids within the bilayer plane, as well as rapid 'flexing' and looser, partially disordered packing of the lipid chains. In liquid-ordered (Lo) phases, the lipid chains exhibit an intermediate degree of order, and the lipids rotate and diffuse in the membrane plane at rates only modestly slower than in the Ld phase.

is relatively smooth and flat, while the opposite side is more irregular. Incorporation of such sterols into bilayers containing phospho- or sphingolipids significantly alters the packing and dynamics of the latter lipid components. Most interestingly, at physiological temperatures, cholesterol forms liquid-ordered phases when mixed with saturated phospho- or sphingolipids but liquid-disordered phases when combined with unsaturated lipids.

Segregation of liquid-ordered and liquid-disordered domains has been observed in bilayers combining saturated phospho- or sphingolipids, unsaturated phospholipids and cholesterol or related sterols [5]. This phenomenon has attracted much attention as a possible model for domain formation in eukaryotic cell membranes, because the plasma membrane, and a variety of intracellular membranes, combine lipids of all three classes just noted. As shown in Figure 2, at temperatures up to 25–30 °C mixtures of lipids from these three classes form coexisting liquid-ordered and liquid-disordered domains that are large enough (>200–250 nm) to be observed by fluorescence microscopy in 'giant' (cell-sized) bilayer vesicles. As the temperature is increased, these domains become smaller, and at 37°C segregated domains typically are no longer observed by fluorescence microscopy. However, fluorescence-spectroscopic approaches, which can detect inhomogeneity in bilayers on shorter distance scales, indicate that at temperatures approaching 37°C, these lipid model systems form smaller nanodomains, with dimensions of nm to tens of nm [6, 7]. These nanodomains may be more relevant to biological membranes than the larger domains formed at lower temperatures, but their precise nature and physical origins remain only partly understood. While detailed discussion of this issue is beyond the scope of this book (for further information see [7]), some current models suggest that nanodomains in these model systems may be not only small but also relatively short-lived (lifetimes of milliseconds to seconds), arising from continuous local fluctuations of lipid composition within the bilayer.

For several lipid mixtures of the type just discussed, comprising a small number of individual lipid components, phase diagrams have been determined that describe, for a given temperature or range of temperatures, the types, compositions, and relative amounts of phases present as a function of the proportions of the different lipid components. The determination and analysis of such phase diagrams is discussed in reference (8). For present purposes, two conclusions from such studies are particularly noteworthy. First, in lipid mixtures exhibiting Ld/Lo phase separation, sterol and saturated lipids are typically only modestly enriched in the Lo phase. Cholesterol, for example, is typically enriched by roughly two-fold in Lo-phase domains compared to coexisting Ld-domains in these systems. Many membrane components described as 'raft-associated' may therefore actually be distributed between non-raft and raft domains, with only a modest enrichment in the latter. Second, for a given 'raft-mimetic' model membrane, Ld/Lo phase separations occur only over a limited range of compositions. A minimum proportion of saturated phospho- or sphingolipid (usually at least 10–20 mol%) must be present in the bilayer, for example, to support formation of segregated

DOPC/DSPC
3:2 (approx.)
+ 35 mol% Cholesterol

DOPC/DSPC
4:5 (approx.)
+ 35 mol% Cholesterol

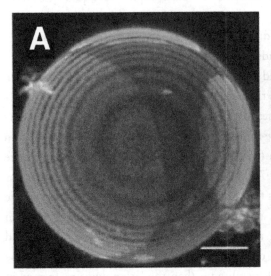
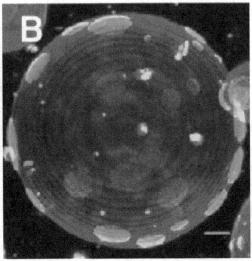

FIGURE 2: Segregated liquid-ordered and liquid-disordered lipid domains can coexist in model membranes. Giant unilamellar vesicles prepared from a mixture of an unsaturated phospholipid (dioleoyl phosphatidylcholine, DOPC), a saturated phospholipid (distearoyl phosphatidylcholine, DSPC), and cholesterol form large coexisting domains of liquid-disordered lipids (green) and liquid-ordered lipids (red) over a range of lipid compositions and temperatures (the vesicles shown were imaged at 23°C). The vesicles shown in the fluorescence micrographs also incorporated low proportions of two fluorescent lipids that preferentially label Ld- and Lo-domains, respectively. As illustrated, the relative proportions (total areas) of the Ld vs. Lo domains vary with the proportions of saturated vs. unsaturated lipids. Space bar = 5 μm (reproduced with permission from Zhao et al., *Biochim. Biophys. Acta* 1768 [2007], 2764–2776).

Lo-phase domains, and very high levels of cholesterol can suppress even nanoscale segregation of domains.

Lipid and lipid–protein model systems have revealed other interesting physical behaviors that may be relevant to the formation of nanodomains in biological membranes. Experiments using 'asymmetric' lipid bilayers, with different lipid compositions in the two leaflets of the bilayer, have revealed that segregation of lipids into liquid-ordered and liquid-disordered domains in one bilayer leaflet can promote segregation of similar domains in the opposite ('trans') leaflet, even when the lipid composition of the *trans* leaflet normally would not favor domain segregation [9, 10]. This result is important because in cellular membranes like the plasma membrane the sphingolipids are localized mainly to the extracytoplasmic leaflet, while the cytoplasmic leaflet contains mainly lipid species that, on their own, do not form segregated liquid-ordered domains in the presence of sterols [11]. Transbilayer coupling of lipid behavior could thus be an important factor in promoting formation of raft domains in the cytoplasmic leaflet of cellular membranes.

Another noteworthy observation from studies of 'raft-mimetic' model membranes is that for some such systems, relatively small perturbations can produce large changes in the overall domain organization of the bilayer. A striking example has been provided by Hammond et al. [12] using giant lipid vesicles composed of a saturated sphingomyelin, an unsaturated phosphatidylcholine, cholesterol and small amounts of a second sphingolipid, ganglioside GM1 (which, like sphingomyelin, favors Lo over Ld-phases). By carefully adjusting the proportions of the different lipid components, these authors created model membranes that were homogeneous under control conditions but exhibited large-scale segregation of Lo- and Ld-phase domains upon addition of the B-subunit of cholera toxin, a pentavalent GM1-binding protein. Examples like these demonstrate that in principle, the organization of lipid domains in cell membranes could be susceptible to large alterations in response to relatively small changes in conditions. It remains to be determined whether cells in fact 'engineer' their membranes to exhibit such high sensitivity to small perturbations, which would presumably require a high degree of regulation to maintain optimal membrane function under varying conditions.

While lipid (and lipid–protein) model membranes have revealed interesting potential mechanisms for nanodomain formation in biological membranes, biological membranes of course exhibit many features that are not satisfactorily replicated in typical model membranes. First, biological membranes are non-equilibrium systems, subject to constant fluxes of materials through both metabolism and trafficking of membrane components, and subjected also to complex, dynamic mechanical perturbations. These features of biological membranes are not satisfactorily reproduced in typical model membrane systems. Second, many biological membranes interact extensively with other cellular structures that may affect membrane structure and dynamics in important ways. The plasma membrane, for example, is intimately associated with the subcortical cytoskeleton through molecular interactions that affect the diffusion and lateral distribution of many plasma membrane components, as discussed more extensively in the book by Kusumi et al. in this series. Finally, membranes contain large amounts of proteins, whose effects on the organization of membrane lipids (particularly in regard to potential domain formation) are in many regards still poorly understood. In the next section, we will consider what is presently known about the interactions of proteins with lipids and how lipid–protein interactions might influence the formation of domains within membranes.

2.2 LIPID–PROTEIN INTERACTIONS

Not surprisingly, intra- and extramembraneous regions of membrane-associated proteins interact with membrane lipids in different ways. Intramembrane regions of proteins (e.g., membrane-spanning helixes) interact extensively, though often with low selectivity, with both the hydrophobic portions of membrane lipids and their polar headgroups. Extramembrane domains of proteins, by contrast, associate mainly with lipid polar headgroups, with widely varying degrees of specificity.

2.2.1 Interactions of Intramembrane Protein Domains with Lipids

High-resolution structures of integral membrane proteins [13, 14] reveal that the lipid-contacting surfaces of these proteins reflect the bilayer organization of the membrane, with a central band of hydrophobic surface exposed to the lipid hydrocarbon chains flanked by more polar regions that interact with the lipid headgroups (Figure 3A). In most eukaryotic cell membranes, the membrane-spanning domains of integral membrane proteins are formed primarily by α-helixes, whose lipid-contacting surfaces comprise mainly aliphatic amino acid residues in the central region of the membrane and include tyrosine and tryptophan residues in the interfacial regions [14, 16]. The lipid-contacting surfaces of integral membrane proteins are irregular and more rigid than lipid hydrocarbon chains. As a result, the hydrophobic chains of lipid molecules in contact with integral membrane proteins exhibit more 'tortuous' conformations (Figure 3A) and less rapid motions than do lipids farther from the protein surface [14]. The conformations of lipid molecules at the protein–lipid interface may also be influenced by mismatch between the lengths of the lipid hydrocarbon chains and the width of the bilayer-spanning hydrophobic surface of the protein, although the magnitude of these effects may be rather modest for physiological combinations of integral membrane proteins and lipids [16].

Most of the membrane lipids associated with integral membrane proteins are associated rather nonspecifically with the protein's lipid-contacting interface, as illustrated in Figure 3B, and

FIGURE 3: 'Boundary lipids' occupy the hydrophobic surfaces of transmembrane proteins. (A) Lipids (purple) imaged at the membrane-exposed surface of water channel-forming protein aquaporin-0. The lipid hydrocarbon chains adopt a variety of conformations to adapt to the irregular surface of the protein. (B) Schematic representation of acetylcholine receptor-rich membrane in the postsynaptic region of a neuronal dendrite, with boundary lipids shown in red. As illustrated in the right panel (a magnified view of the region indicated by the arrow in the left-hand panel), even non-'boundary' lipids in membranes typically lie at most a few nanometers (the center-to-center distance between the lipids shown is ca. 0.85 nm) from the nearest transmembrane protein molecule. ((A) adapted with permission from reference [21]; (B) reproduced with permission from Barrantes, *Brain Res. Rev.* 47 [2004], 71–95).

A

Extracytoplasmic Face

PC1 PC2 PC3 PC4

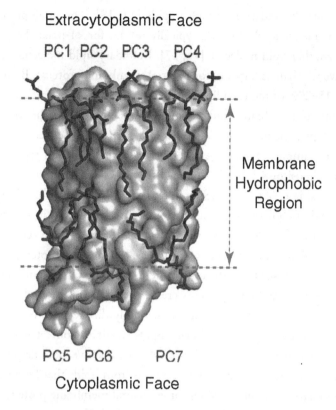

Membrane
Hydrophobic
Region

PC5   PC6   PC7

Cytoplasmic Face

B

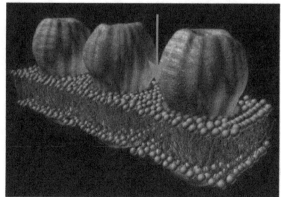 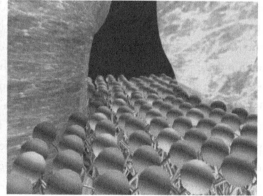

have been termed boundary or annular lipids [17, 18]. The lipid–protein interface is extremely dynamic; a boundary lipid molecule typically resides for <1 µs at the protein surface before it is replaced by another lipid molecule [16, 17]. Boundary lipid molecules can therefore be regarded more as 'solvents' than as 'ligands' for integral membrane proteins. Boundary lipids are estimated to comprise 15–20% of the total lipid population in a typical membrane at any instant [19, 20]. A smaller fraction of membrane lipids is bound to membrane proteins with higher specificity, in the manner of a typical ligand or cofactor, sometimes in clefts between transmembrane helixes [18]. A number of examples of such specific lipid binding have been reported, including the binding of phosphatidylinositol-4,5-bisphosphate (PIP_2) to GIRK2 potassium channels, of cardiolipin to cytochrome c oxidase, of C18-acylceramide to a p24 protein (involved in ER-to-Golgi membrane trafficking), and of cholesterol to the amyloid precursor protein implicated in the pathogenesis of Alzheimer's disease [21–24]. Such specifically bound lipids can be very important in regulating the function or stabilizing the structure of an integral membrane protein, even though they generally comprise only a small subset of the total protein-associated lipid molecules.

Can integral membrane proteins create nanodomains around themselves, with lipid compositions and physical properties that are markedly different from those found elsewhere in the membrane? If we consider the boundary lipids in direct contact with a protein to constitute a 'nanodomain,' the answer appears to be yes. As noted above, boundary lipids differ significantly in their dynamics and average chain conformations from lipids that lie further from the protein surface. The boundary lipids surrounding some integral membrane proteins can also differ somewhat in composition from the overall membrane lipid pool. For example, anionic lipids are somewhat enriched in the boundary lipid layer of the (Na^+,K^+)-ATPase (relative to the membrane overall), attracted by an excess of positively over negatively charged amino acid residues at the lipid–protein interface [17]. Some integral membrane proteins show a degree of selectivity in their interactions with membrane lipids based on the length of the lipid hydrocarbon chains [18]. However, in the examples studied to date, the degree of enrichment (or depletion) of particular phospho- or sphingo-lipid species in the boundary lipid layer relative to the membrane overall is typically modest (less than twofold). Some membrane proteins, like the Ca^{2+}-ATPase of sarcoplasmic reticulum, appear to show stronger discrimination against sterols in their boundary layer, possibly because the relatively rigid sterol nucleus accommodates poorly to the similarly rigid protein surface. Integral membrane proteins thus can modestly adjust the composition, as well as influencing the dynamics and conformations, of their boundary lipids. By contrast, the perturbing effects of integral membrane proteins on lipid conformation and dynamics appear to be much smaller for lipids beyond the boundary layer. There is little biophysical evidence to date that integral membrane proteins, at least in a monomeric (unclustered) form, can organize around themselves lipid nanodomains, with lipid compositions and physical properties markedly different from those of the surrounding bilayer, that extend beyond the boundary-lipid layer itself. There is however a need for detailed biophysical studies of the lipid interactions of additional membrane proteins, particularly proteins proposed to associate with 'raft' lipids, to assess this question more fully.

2.2.2   Interactions of Extramembranous Proteins and Protein Domains with Lipids

Membrane proteins can also associate with lipids at the membrane surface, interacting principally or even exclusively with the lipid polar headgroups. As illustrated in Figure 4A, some such interactions are mediated through well-structured protein binding sites that recognize particular lipid headgroups with high selectivity and a defined stoichiometry, as for the various PH domains that bind different phosphoinositides or the domains of bacterial toxins that bind specific glycosphingolipids [25, 26]. Since the interactions of these protein domains with the bilayer involve only one or a very few lipids, such interactions may seem unlikely to create lipid domains of significant size. However, as discussed in Section 2.1, it has been shown that the B-subunit of cholera toxin, a pentameric protein that binds five molecules of ganglioside GM1, can promote major changes in the lateral organization of lipids in suitably engineered model membranes [12], illustrating the possibility that multivalent binding of proteins to specific lipid headgroups could contribute to formation of membrane domains substantially larger than the dimensions of individual protein molecules.

Another type of protein–lipid interaction at the membrane surface, based on electrostatic attraction between anionic lipids and positively charged regions on protein surfaces, could also contribute significantly to formation of membrane domains. Regions of proteins that carry a high net positive charge and are closely apposed to the membrane surface can promote enrichment of anionic lipids in the area of protein–membrane contact, as has been demonstrated for the MARCKS protein which attracts and sequesters phosphatidylinositol-4,5-bisphosphate (PIP_2) through a membrane-associated polybasic region (Figure 4B and [27]). Theory and experiment both suggest that such 'clustering' effects will be relatively weak for monoanionic lipids such as phosphatidylserine or phosphatidylinositol but can be considerably stronger for lipids with a high net negative charge, such as PIP_2 or phosphatidylinositol-3,4,5-trisphosphate [28].

2.2.3   Determinants of Protein Association with Liquid-Ordered vs. Liquid-Disordered Lipids

Studies with lipid and lipid–protein model systems have provided useful insights into some aspects of lipid–protein interactions that may influence protein association with raft or nonraft domains in

A

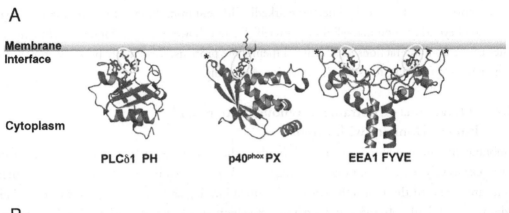

B

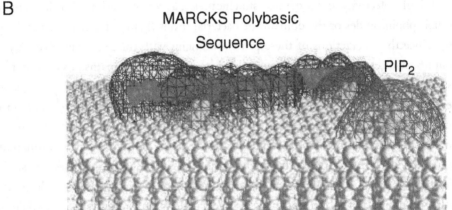

FIGURE 4: Peripheral membrane proteins can interact in different manners with lipid polar head-groups. (A) Specialized domains found in many peripheral membrane proteins bind individual lipid headgroups with high specificity. The illustrated PH, PX, and FYVE domains, for example, bind specific phosphoinositides, whose polar headgroups are shown within the yellow ovals (the lipid acyl chains are largely omitted for clarity). (B) Some peripheral membrane proteins possess strongly positive charged sequences that cluster polyanionic lipids on the basis of their net charge, not their detailed structure. A strongly cationic (net charge +13) membrane-binding sequence from the MARCKS protein, for example, creates a strongly positive local electric field (blue) that attracts polyanionic membrane lipids like phosphatidylinositol-4,5-bisphosphate (net charge −4 to −5), which create strongly negative local electric fields (red). ((A) Adapted with permission from Moravcevic et al., *Structure* 20 [2012], 15–27; (B) reproduced with permission from McLaughlin and Murray, *Nature* 438 [2005], 605–611).

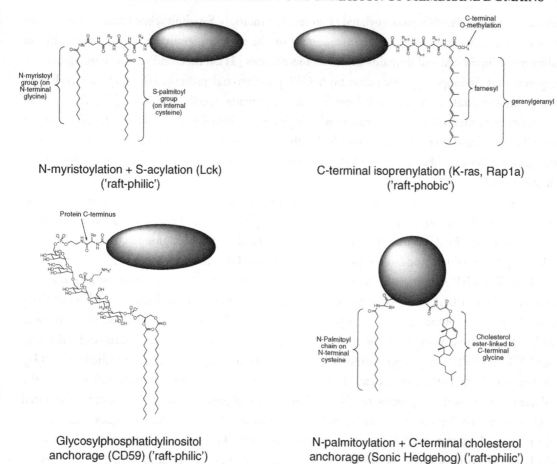

N-myristoylation + S-acylation (Lck)
('raft-philic')

C-terminal isoprenylation (K-ras, Rap1a)
('raft-phobic')

Glycosylphosphatidylinositol
anchorage (CD59) ('raft-philic')

N-palmitoylation + C-terminal cholesterol
anchorage (Sonic Hedgehog) ('raft-philic')

FIGURE 5: Structures of some common lipid modifications of proteins. N-myristoylation, S-acylation (often termed 'palmitoylation'), and isoprenylation are found on cytoplasmic proteins and protein domains. By contrast, GPI-anchorage and the dual N-acyl/cholesterol modifications of Sonic Hedgehog and related proteins are found on extracytoplasmically localized proteins. The structure shown for a GPI-anchor is representative, but the 'anchor' structure can vary between different GPI-proteins, for example incorporating additional glycosyl residues.

biological membranes. Glycosylphosphatidylinositol (GPI-) anchors (illustrated in Figure 5), which in their mature form typically carry two long saturated hydrocarbon chains, as individual molecules exhibit a modest but significant affinity for liquid-ordered lipid domains [29, 30]. Interestingly, the affinity of GPI-anchored proteins for Lo domains in model systems is enhanced when they are artificially clustered [29, 30]. Protein sequences modified with multiple saturated acyl chains, such as the N-myristoylated/S-palmitoylated (Figure 5) or multiply S-palmitoylated motifs that anchor a number of proteins to the cytoplasmic surface of the plasma membrane, also show significant affinity for liquid-ordered domains in model membranes [31]. These findings are consistent with experimental findings suggesting that both GPI-proteins and proteins anchored to membranes via multiple saturated acyl chains behave as raft components in cellular systems [32]. By contrast, isoprenylated protein sequences partition with high affinity into liquid-disordered domains in lipid model systems, and even sequences modified with both an isoprenyl group and a saturated acyl residue (as found in a number of isoprenylated proteins) show a strong preference for liquid-disordered domains [31].

An obvious means to target peripheral membrane proteins to liquid-ordered lipid domains is via protein binding to the polar headgroups of 'raft-philic' lipids. Two well-studied examples of such proteins are the B subunit of cholera toxin, which binds to the headgroup of ganglioside GM1, and Shiga toxin, which binds to the globotriaosylceramide Gb3, in both cases in a multivalent manner [26]. The HIV Gag protein interacts with its membrane lipid target, phosphatidylinositol-4, 5-bisphosphate (PIP_2), in a more complex manner [33]. As illustrated in Figure 6, HIV Gag binds both the polar headgroup of PIP_2 and the polyunsaturated 2-position acyl chain at its 2-position, which the protein extracts from the bilayer and sequesters in a hydrophobic binding cleft. Binding of HIV Gag to PIP_2 also triggers exposure of the N-terminal myristoyl group attached to the Gag protein. The Gag/PIP_2 complex can then associate with the membrane by intercalating into the bilayer two saturated acyl groups, the N-terminal myristoyl group of the protein and the saturated acyl chain at the 1-position of the bound PIP2 molecule (Figure 6), while sequestering within the protein, the unsaturated 2-position acyl chain of PIP_2. This mode of interaction would allow the Gag-PIP_2 complex to associate with liquid-ordered domains without incurring the high energetic penalty that would accompany insertion of the polyunsaturated PIP_2 acyl chain into a liquid-ordered bilayer [33].

Model-system studies to date have provided fewer clear insights into the mechanisms by which some transmembrane proteins may associate with raft domains in biological membranes. Artificial bilayer-spanning peptides with simple model sequences, and an S-palmitoylated peptide corresponding to the transmembrane sequence of LAT, an adaptor protein involved in immune-cell signaling, have been found to be effectively excluded from liquid-ordered domains in mixed-phase Lo/Ld bilayers [34, 35]. The latter result was somewhat surprising, since LAT appears to behave as a raft-associating protein in biological systems. Two explanations have been suggested for these seemingly divergent findings regarding LAT. First, protein–protein interactions mediated by the cytoplasmic domain of LAT are important for the association of this protein with putative 'raft' fractions isolated biochemically from cell membranes, suggesting that these protein–protein in-

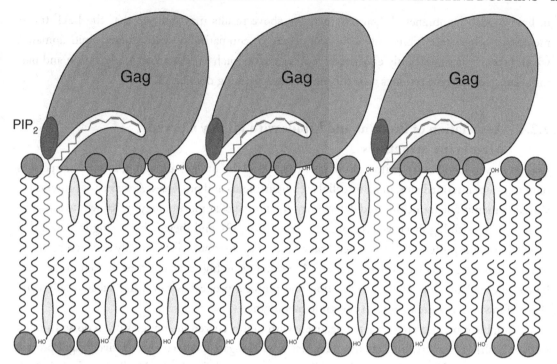

FIGURE 6: The HIV Gag protein binds to membranes in a distinctive manner that may facilitate its association with lipid-ordered domains. The 2-position acyl chain of phosphatidylinositol-4,5-bisphosphate (PIP$_2$, shown in red) is polyunsaturated and is not well accommodated in a liquid-ordered lipid environment. The Gag protein binds both to the polar headgroup and the 2-position acyl chain of PIP$_2$, sequestering the polyunsaturated chain inside the protein interior. The saturated acyl chain at the 1-position of PIP$_2$ intercalates in the bilayer along with the saturated myristoyl chain (cyan) attached to the Gag N-terminus, thereby anchoring PIP$_2$-bound Gag proteins (which can cluster, as shown) to the bilayer through two 'raft-philic' acyl chains.

teractions may enhance the ability of the transmembrane domain to associate with ordered lipids [35]. Second, in blebbed plasma membrane vesicles, in which coexisting liquid-ordered and liquid-disordered domains can be observed by fluorescence microscopy (see Section 5.10), the LAT membrane-spanning sequence can under some conditions associate with ordered-lipid domains [36]. It has been suggested that in biological membranes, with their high protein contents and very diverse lipid compositions, ordered-lipid domains may be less highly ordered, and more accommodating to structures like the LAT transmembrane domain, than are the liquid—ordered domains formed in lipid model membranes. Taken together, the above results may indicate that the LAT transmembrane sequence exhibits only a limited degree of compatibility with ordered-lipid domains, which however is significantly enhanced by other factors such as protein oligomerization and palmitoylation of cysteine residues near the membrane-spanning domain [37].

2.2.4   Association of Proteins and Lipids with Highly Curved Membrane Structures

Highly curved membranes (with radii of curvature of 50 nm or less) are found, as stable structures or as transient intermediates, in many cellular compartments and processes. A variety of proteins involved in membrane trafficking and determination of organellar morphology are known to associate preferentially with highly curved regions of membranes [38], and there is increasing evidence that other membrane proteins and even membrane lipids may distribute differentially between high- *vs.* low-curvature regions of a given membrane. Accordingly, highly curved regions of membranes may be considered 'domains' whose composition and physical properties can differ substantially from those of lower-curvature regions of the same membrane.

Soluble proteins that associate selectively with highly curved membranes often do so through one (or both) of two structural motifs. The first is an elongated, curved and relatively rigid protein face that contacts the membrane lipid surface over an extended distance, binding preferentially to membrane regions where the local radius of curvature closely mirrors that of the lipid-interacting face of the protein (Figure 7A). These motifs are exemplified by the so-called BAR domains, found in proteins that play diverse roles in membrane trafficking and other cellular processes [38–40]. Proteins of a second group bind preferentially to highly curved membranes by shallow insertion of the apolar faces of amphiphilic α-helixes into the lipid bilayer, an interaction that is favored by increasing (positive) curvature of the membrane surface (Figure 7B). The curvature-promoting abilities of both types of proteins, and their differential distribution between membrane regions of high and low surface curvature, can be enhanced through oligomerization at the membrane surface.

Integral membrane proteins can also dictate and/or sense membrane curvature, as in the case of the reticulons and DP1/Yop1p-family proteins found in tubular ER [41]. The membrane tubule-forming properties of these proteins may rest on a combination of the shape of the protein monomers (reticulons and DP1/Yop1p proteins, for example, have been proposed to occupy a larger cross-sectional area in the cytoplasmic than in the luminal leaflet of the ER membrane) and assembly of these proteins into oligomers with a rigid geometry that promotes membrane bending, as illustrated in Figure 7C.

Proteins like those discussed above, which are 'professionally' engaged in sensing and promoting membrane curvature, may not be the only membrane components that distribute differentially between regions of high and low curvature within a given membrane. GPI-proteins and other

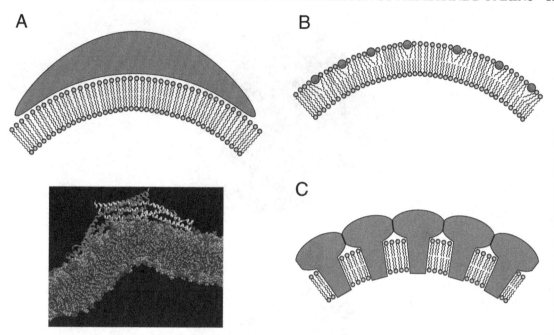

FIGURE 7: Membrane-associated proteins can sense or promote membrane curvature through diverse structural features. (A) A BAR domain binds to lipid polar headgroups with selectivity for regions where the curvature of the membrane complements that of the BAR domain's lipid-binding surface. (B) Amphiphilic helixes (shown 'end-on' in cross-section) insert their hydrophobic faces (red) into the outer region of the bilayer's hydrophobic interior, expanding the membrane leaflet to which they bind. Through such intercalation, proteins can 'sense' or even promote bilayer curvature. (C) Integral membrane proteins may promote (or sense) bilayer curvature by at least two mechanisms. First, proteins that occupy a larger cross-sectional area in one bilayer leaflet than in the other will favor local membrane curvature. Second, integral membrane proteins that oligomerize into a rigid, curved lattice can 'scaffold' local membrane curvature, even at much lower densities within the membrane than in the schematic illustration shown here. (Lower panel in (A) reproduced with permission from Blood and Voth, *Proc. Natl. Acad. Sci. USA* 103 [2006], 15068–15072.)

highly 'asymmetric' membrane proteins, for example, which concentrate most or all of their mass on one side of a membrane, have been shown experimentally [42, 43] to favor membrane regions with appropriate curvature, which reduce steric repulsions with other membrane proteins (Figure 8). Other membrane proteins may show varying distributions between membrane regions of different curvature, based on their distributions of mass across (or within) the bilayer or the geometry of the oligomers they form, in accord with the physical principles discussed above.

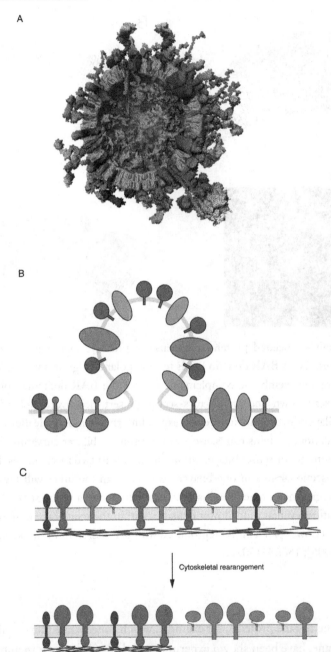

FIGURE 8: Steric crowding can influence the behavior of membrane proteins in unexpected ways. (A) The high density of proteins in most biological membranes is illustrated by this reconstruction of the membrane of a synaptic vesicle, based on careful compositional analysis. The 'crowding' of membrane proteins, many of which are laterally mobile, leads to frequent collisions and strong (repulsive) steric inter-

actions. (B) Membrane bending increases protein crowding at one face of the membrane while reducing crowding at the other. 'Asymmetric' proteins, whose steric 'footprints' (effective cross-sectional areas) differ markedly at the two faces of the membrane, can become enriched or depleted in highly curved regions of the membrane in response to these effects, as illustrated. (C) Modulating directly the distribution of one class of membrane proteins can influence the distributions of other proteins indirectly, through steric-crowding effects. In the hypothetical example shown, redistributing one class of abundant membrane proteins (red/violet), by rearrangement of the juxtamembrane cytoskeleton, will induce compensatory redistribution of other (mobile) membrane proteins (blue/green) in order to minimize globally the free energy of steric repulsions between proteins. ((A) reproduced with permission from reference [19]).

The potential for lipids to be 'sorted' (i.e., show differential partitioning) between high- and low-curvature regions of membranes has been assessed using giant unilamellar lipid vesicles from which highly curved tubules, still connected to the parent vesicle, can be pulled using mechanical force (Figure 9). When the vesicle lipids form a single homogeneous fluid phase, different lipids partition equally between the vesicle body and the attached tubule, even when the radius of curvature of the tubules is as small as 15 nm [44]. However, in similar experiments using tubules pulled from vesicles that exhibit liquid-ordered/liquid-disordered (Lo/Ld) phase separation, the tubule membrane is significantly depleted, compared to the 'parent' vesicle, in the lipid components of the Lo phase [44, 45]. The lipid sorting observed in the latter system appears to reflect a collective sorting behavior of the lipids, based on a greater resistance of Lo than of Ld lipid domains to incorporation into highly curved membranes. Interestingly, similar lipid sorting can also be observed in tubules pulled from vesicles whose lipid compositions lie *close to*, but not within, the range of compositions that supports segregation of optically detectable Ld and Lo domains in the parent vesicle [46]. This behavior is particularly noteworthy in the light of suggestions that some cellular membranes, at least in their 'resting' condition, may exist in a state that thermodynamically lies close to, but not actually within, a regime of lipid phase separation.

2.3   PROTEIN–PROTEIN INTERACTIONS IN MEMBRANES

The great diversity and structural complexity of membrane-associated proteins allows an enormous potential range of protein–protein interactions within membranes, our understanding of which is still quite limited. However, we will review here some general principles that can be particularly important for understanding the interactions among membrane proteins and their potential relevance to the formation of membrane domains.

Incorporating a protein into (or onto) a given cellular membrane confines the protein to a very small subspace within the total cellular volume, restricts its spatial orientation, and constrains

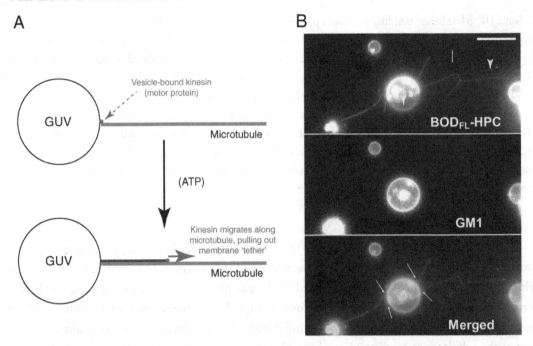

FIGURE 9: Liquid-ordered and liquid-disordered lipids distribute differentially between membrane regions of high *vs.* low curvature. (A) Narrow, highly curved bilayer tubules ('tethers'), continuous with the bilayer of a giant unilamellar vesicle (GUV), can be pulled from the GUV in vitro by vesicle membrane-anchored kinesins moving along added microtubules (for further experimental details see [47]). (B) A fluorescent phosphatidylcholine (BOD$_{FL}$-HPC) that selectively associates with liquid-disordered domains (top image) partitions readily into highly curved 'tethers' (yellow arrows), which accordingly are visibly labeled with this marker. At the same time, a marker for liquid-ordered domains (middle image = fluorescent cholera toxin B-subunit bound to bilayer-incorporated ganglioside GM1), is excluded from the membrane tethers. Space bar = 10 μm. Panel B reproduced with permission from reference (45).

the vertical positions of different parts of the protein vis-à-vis the membrane surfaces (Figure 10). All of these factors can promote protein–protein interactions within membranes that, while significant (and biologically important) in the intact membrane, may be easily disrupted if the membrane itself is solubilized by detergents as is common in biochemical procedures. Many peripheral membrane proteins associate with membranes by forming multiple simultaneous, individually weak interactions with different membrane lipid and/or protein molecules; again, such interactions are readily abolished when the membrane is disrupted. Many interactions between membrane proteins therefore can be satisfactorily characterized only by studying them in the intact membrane.

Proteins in solution:
- Randomly oriented about x, y and z-axes
- Randomly positioned in x, y and z-directions

Proteins anchored to membrane:
- Highly concentrated in membrane plane
- Orientation about x- and y-axes fixed
- z-position fixed relative to other proteins

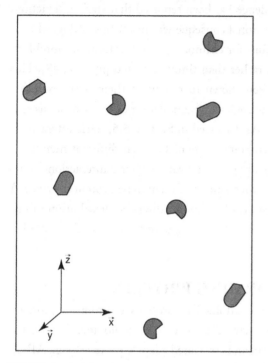 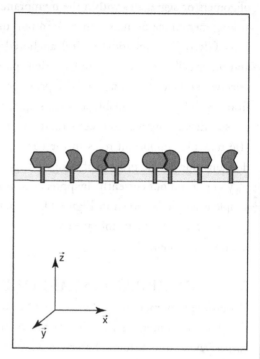

FIGURE 10: Membrane anchorage promotes protein-protein interactions in multiple ways. Localizing a given set of proteins to a membrane concentrates the proteins within a very small fraction of the total cellular volume, restricts the orientations of the proteins about two principal axes of space (**x** and **y** in the illustration) and constrains their relative positions along the third principal axis (**z**), as shown. All of these factors can promote protein-protein interactions. Membrane-bound proteins can therefore associate through interactions that would be too weak to be significant if the proteins were free in the aqueous phase.

Three useful methods to study protein–protein interactions in living cells are fluorescence resonance energy transfer (FRET), bimolecular fluorescence complementation (BiFC), and single-molecule fluorescence analyses of the interactions of individual protein molecules. These methods and their potential (and limitations) for studying domain formation in membranes are discussed further in Sections 5.6–5.8.

In a typical biological membrane, proteins occupy a large fraction of the membrane surface area [19, 20], creating steric 'crowding' as proteins undergo frequent collisions. Crowding effects, while relatively nonspecific, can substantially influence a variety of aspects of the behavior of membrane proteins, including slowing their lateral diffusion and enhancing their tendencies to form oligomers or aggregates within the membrane. Evidence has been reported that steric interactions among membrane proteins can have important functional consequences in various biological contexts. Glycosylphosphatidylinositol-anchored proteins, for example, are endocytosed in fibroblasts and other cells by a clathrin-independent pathway rather than through coated pits [47, 48]. This selective endocytic routing of GPI-proteins has been shown to rest upon their steric exclusion from coated pits, presumably as a consequence of dense local accumulation of other proteins that are specifically targeted to these structures [49, 50]. As discussed in Section 5.5, steric interactions also may play important roles in the sorting of membrane proteins between different membrane domains during the process of antigen-triggered lymphocyte activation. Steric interactions could play important but currently unappreciated roles in other aspects of membrane protein behavior. A simple example is shown in Figure 8C, which illustrates how modulating the distribution of one group of proteins in a membrane could induce complementary changes in the distributions of other membrane proteins.

2.4 MEMBRANE 'SCAFFOLD'-FORMING PROTEINS

Various types of membrane domains are known to be 'scaffolded' by proteins, or groups of proteins, that form oligomeric assemblies within or immediately adjacent to the membrane. Among the earliest such proteins to be characterized were clathrin and associated adaptor proteins (e.g., AP-2), which among other functions participate in the formation of endocytic coated pits and vesicles. A number of other coat-forming proteins and their associated adaptors, implicated in formation of a variety of different types of transport vesicles, have since been characterized. Caveolins, assisted by proteins such as PRTF/cavin and pacsins/syndapins, oligomerize to form a morphologically well-defined scaffold for caveolae. Several types of cell–cell adhesions, including the zonula occludens (tight junctions), zonula adherens, and desmosomes, are built on fairly stable and spatially well-differentiated microdomains of integral membrane proteins, such as claudins, occludin, and cadherins, linked to juxtamembrane proteins and, in some cases, to the adjacent cytoskeleton. These structures and their constituent proteins will not be discussed further in this book, but further information can be found in the books by Stan and by Goetz in this series. As discussed in Section 2.2.4, integral and peripheral membrane proteins have also recently been identified whose lipid-binding properties, intrinsic 'shapes' and/or geometry of oligomerization play key roles in forming or stabilizing morphologically differentiated features of membrane-bound compartments like the ER, which comprises both tubular and cisternal elements [51, 52], and in generating highly curved membrane regions during processes such as membrane trafficking.

The above examples illustrate how integral and peripheral membrane proteins, in some cases working together with the juxtamembrane cytoskeleton (and/or the extracellular matrix), create a variety of well-characterized domains within cellular membranes. There are ample indications that proteins may also help to 'scaffold' various types of membrane nanodomains that are smaller, shorter-lived or less spatially well defined, and hence at present less well understood, than the types of membrane domains discussed above. Some of these types of proteins are described below.

Tetraspanins are a family of integral membrane proteins that interact both with other tetraspanin molecules and with a number of membrane-associated partner proteins, including diverse integrins and other adhesion proteins, to form extensive networks of interacting proteins sometimes referred to as a 'tetraspanin web' [53]. A specialized (and unusually well defined) example of a tetraspanin-containing protein network is found on the apical surface of cells of the urothelium, the epithelium that lines much of the urinary tract. The apical plasma membrane of these cells is largely occupied by 24-meric complexes, composed of six molecules each of two different tetraspanins and two associated partner proteins, which assemble into a two-dimensional quasi-crystalline array. The structure of this array has been determined to 6 Å resolution [5], revealing an extensive repeating network of interactions both between adjacent tetraspanin molecules and between tetraspanins and their associated partner proteins (Figure 11). Biophysical evidence indicates that in other types of cells tetraspanins (and their partner proteins) may associate through similar interactions but in a more dynamic, less rigidly defined manner [55]. Tetraspanins may promote and organize the clustering of their 'partner' proteins, both through direct interactions within the membrane and through indirect interactions with the submembrane actin cytoskeleton [53].

Integral membrane proteins known as flotillins-1 and -2, as well as the MAL and MAL2 proteins, have been functionally linked to trafficking of raft components in a variety of cell types and are largely isolated in DRM fractions, consistent with the possibility that they may associate preferentially with rafts and/or promote raft formation or enlargement [56–58]. Flotillins resemble caveolins in their predicted transmembrane topologies (though not in their sequences) and, like caveolins, readily form homo- and hetero-oligomers. When co-expressed, flotillin-1 and -2 form punctate structures in the plasma membrane, which are distinct from caveolae [56]. MAL and MAL2 both possess four membrane-spanning helixes; MAL but not MAL2 has been found to form isolable oligomers [59]. Further elucidation of the tertiary/quaternary structures and the lipid interactions of both of the above classes of proteins may help to clarify the bases for their apparent association with rafts and their contributions to trafficking of raft components.

While caveolae are more stable and morphologically better defined than the types of nanodomains that form the main focus of this book, a relationship between caveolae and membrane rafts has long been suggested, based on various lines of experimental evidence. In cells expressing caveolins these proteins are recovered along with typical raft components in low-density detergent-resistant membrane (DRM) fractions. Depletion of membrane cholesterol moreover leads

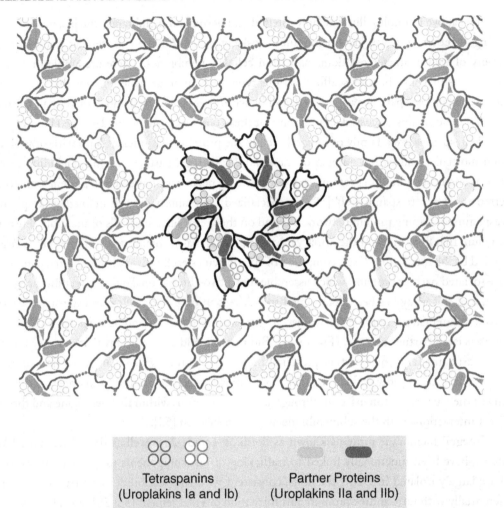

Tetraspanins
(Uroplakins Ia and Ib)

Partner Proteins
(Uroplakins IIa and IIb)

FIGURE 11: Tetraspanins and partner proteins form an extensive, highly ordered 'tetraspanin web' in the plasma membrane of urothelial cells. The structure shown, determined by electron diffraction (resolution = 6 Å) from purified urothelial cell plasma membrane fragments, reveals a highly ordered two-dimensional intramembrane protein lattice, formed by the tetraspanins uroplakin 1a and 1b and their (non-tetraspanin) partner proteins uroplakin IIa and IIb. As shown, the basic unit of the lattice (irregular shapes outlined) comprises one molecule each of uroplakins 1a, 1b, IIa and IIb (tetraspanin monomers are shown as quartets of circles representing the four membrane-spanning helixes). Red arrows/lines indicate the three distinct types of protein–protein interactions through which these fundamental units associate to form hexamers that in turn assemble into an extended hexagonal lattice (reproduced with permission from reference [54]).

to disappearance of morphologically detectable caveolae, suggesting a further potential similarity to rafts [60]. However, a variety of evidence also indicates that caveolae are not simply 'caveolin-scaffolded rafts.' While some early studies suggested that GPI-proteins, a 'classical' raft marker, were concentrated in caveolae, later work found that this localization was an artifact of the visualization conditions and that GPI-proteins are normally not enriched in these structures [61, 62]. As well, biochemical isolation of caveolae using a detergent-free, morphology-based approach showed that the protein composition of caveolae is very different from that of rafts, undoubtedly reflecting at least in part the interactions of a variety of membrane proteins with the scaffolding domain of caveolin-1 [62, 63]. Thus while caveolae exhibit some properties similar to those of rafts (e.g., sensitivity to membrane cholesterol levels), caveolins and associated proteins clearly play a major role in shaping the composition and other properties of caveolae. The same may be true for membrane nanodomains 'scaffolded' by, e.g., flotillins, a non-caveolar pool of plasma membrane caveolin [64] or MAL proteins.

Another class of proteins that may play important roles in formation of membrane nanodomains is the family of carbohydrate-binding proteins known as galectins. These proteins typically bind to carbohydrate determinants based on an N-acetyllactosamine core, although different galectins bind optimally to different specific carbohydrate structures. Galectins can bind glycosylated molecules in a multivalent manner, either through tandem carbohydrate-recognition domains in a single polypeptide chain (as in galectin-4, -8 and -9) or by forming oligomers (as for galectin-1 and -3). They can thus promote formation of clusters or 'lattices' of multiply glycosylated molecules or molecular aggregates, as illustrated in Figure 12. Galectins lack classical leader sequences for translocation into the endoplasmic reticulum and are partly found in the cytoplasm. However, they can also undergo translocation from the cytoplasm to reach the cell surface, by a nonclassical pathway whose subcellular localization has not been fully defined but appears to involve neither the ER nor the Golgi. A fraction of total cell-associated galectins is therefore found associated with glycosylated molecules on the cell surface.

The (partial) cell-surface localization of galectins, and their potential to crosslink multiply glycosylated molecules into extensive assemblies, have led to proposals that galectins form domains or 'lattices' on cell surfaces, into which glycoproteins (and glycolipids) bearing appropriate carbohydrate determinants could be recruited [65, 66]. Evidence supporting this proposal has been reported from studies using cells deficient in the *Mgat5* gene product, a glycosyltransferase essential for synthesis of the oligosaccharide groups to which galectins bind, or treatments with competitive inhibitors to disrupt galectin-glycoprotein interactions on cell surfaces [65]. Association of a glycoprotein with a galectin-generated lattice on the cell surface (or in another cellular compartment) could influence the glycoprotein's distribution and dynamics, and thereby its function. Inhibiting

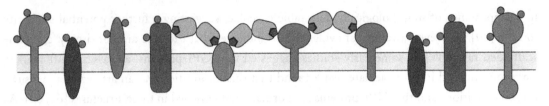

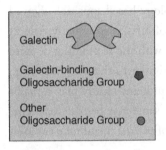

FIGURE 12: 'Galectin lattices' can link glycoproteins dynamically into clusters or networks on the cell surface. Bi- or multivalent galectins can bridge glycoproteins bearing appropriate oligosaccharide determinants, forming dynamic networks if the glycoproteins carry multiple oligosaccharide residues of the appropriate type.

the interactions of the epidermal growth factor (EGF) receptor with galectin lattices on breast carcinoma cells, for example, enhances association of the receptor with plasma membrane-associated caveolin and thereby reduces receptor activity [67]. Galectin/glycoprotein interactions can also regulate the trafficking of membrane proteins such as the EGF receptor, whose constitutive internalization is negatively regulated by association of the receptor with the galectin lattice on the cell surface [68]. Galectins may also function to promote transport of some glycosylated proteins, for example by helping to recruit such proteins into nascent transport vesicles.

•   •   •   •

CHAPTER 3

Kinetic Bases for Formation of Membrane Domains

A second class of possible mechanisms for 'domain' formation in membranes rests on a kinetic rather than a thermodynamic basis. Particular membrane components may be concentrated (or depleted) in particular regions of the membrane by active, time-dependent processes including intra- and intermembrane transport, synthesis, degradation and reversible chemical modifications. As the examples below illustrate, there are experimentally established precedents for multiple kinetically based mechanisms of domain formation.

Epithelial cells maintain separate apical and basolateral domains within their plasma membranes, with a boundary at the level of the intracellular tight junctions (Figure 13A). Establishment and maintenance of these domains rests on two elements: first, barriers that impede lateral diffusion of many membrane proteins and lipids across the region of the intercellular tight junctions, and second, trafficking pathways that selectively deliver distinct complements of membrane proteins and lipids to the apical vs. basolateral regions of the plasma membrane. Similar mechanisms may operate in the formation of less long-lived, but also spatially extensive (μm-sized) domains in the plasma membranes of cells such as macrophages and budding yeast. In macrophages, diffusion barriers restrict free lateral movement of proteins and lipids between a region of the plasma membrane known as the phagocytic cup, with which the macrophage gradually surrounds and ultimately engulfs bacteria and other 'target' particles, and the remainder of the plasma membrane [69]. These diffusion barriers work together with both exocytic trafficking of membrane materials [70] and remodeling of lipid components within the phagocytic cup in order to differentiate the composition of the plasma membrane in the cup region from that of other regions of the membrane. While the nature of these diffusion barriers has yet to be defined, it has been suggested they may be formed at least in part by juxtamembrane molecules known as septins, which have been shown to form barriers restricting the lateral diffusion of membrane proteins in budding yeast [71]. Septins also form part of a diffusion barrier implicated in formation of a specialized plasma membrane domain that envelops the primary cilium in mammalian cells [72, 73]. It will be interesting to determine whether similar diffusion barriers contribute to the formation of other types of membrane nanodomains.

A

Polarized Renal or
Intestinal Epithelial Cells

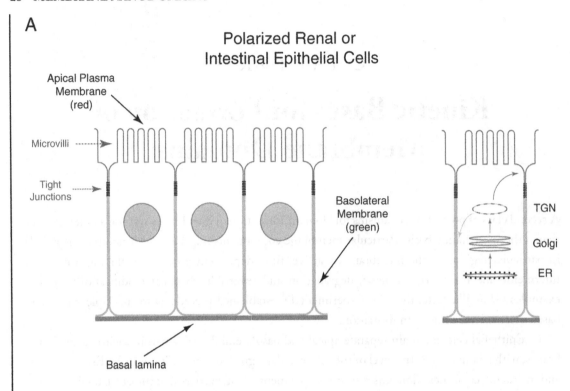

Apical Plasma
Membrane
(red)

Microvilli

Tight
Junctions

Basolateral
Membrane
(green)

Basal lamina

TGN

Golgi

ER

B

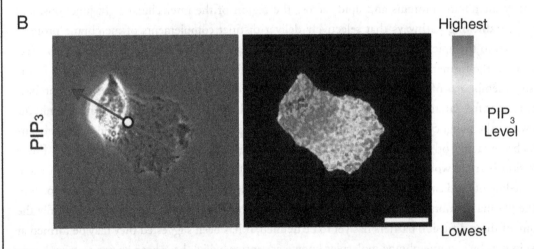

PIP$_3$

Highest

PIP$_3$
Level

Lowest

A less sharply demarcated type of 'domain' can be created in a membrane, again through a kinetic mechanism but in this case without requiring specific diffusion barriers, by generating a spatial gradient of particular membrane components. Gradients can be formed by introducing a particular membrane component selectively in one region of the membrane while at the same time gradually removing it (e.g., by degradation or membrane traffic) from other regions. *D. discoideum* and human neutrophils undergoing chemotaxis, for example, selectively generate phosphatidylinositol-3, 4,5-trisphosphate (PIP_3) at the leading edge of the cell while degrading this lipid in other regions of the membrane, thereby producing a gradient of PIP_3 polarized toward the leading edge (Figure 13B and [74, 75]). Gradients of this type can be created only on spatial scales over which the rate of lateral diffusion of the species of interest is slower than the rate of removal of that species from the membrane. For membrane lipids, which diffuse rather rapidly in membranes, concentration gradients formed by this mechanism typically extend over relatively long length scales (several μm or more).

Many membrane-associated proteins diffuse more slowly than membrane lipids, for example as a consequence of transient obstruction by juxtamembrane structures like actin microfilaments (see the book by Kusumi et al. in this series). For such proteins, concentration gradients may be generated within a membrane over smaller (even submicron) spatial scales. On the surface of murine fibroblasts, for example, transient domains enriched in the major histocompatibility complex-1 (MHC-1) have been observed with dimensions of hundreds of nm and average lifetimes of 20–30 s. Modeling and experimental data suggest that these transient domains are formed by fusion of MHC-1-enriched exocytic vesicles with the plasma membrane, and that the domains subsequently

---

FIGURE 13: Large-scale 'domains' can be created by multiple mechanisms in cell plasma membranes. (A) In epithelial cells, stable diffusion barriers at the level of tight junctions divide the plasma membrane into sharply differentiated apical and basolateral domains. As shown at the right, the secretory pathway delivers different proteins and lipids from the trans-Golgi network to the apical versus the basolateral domains. Endocytic pathways (not shown) are also partly segregated into apical and basolateral 'limbs' to maintain the distinct compositions of these membrane domains. (B) More diffuse membrane 'domains,' enriched in particular lipids or proteins, can be created in membranes even in the absence of diffusion barriers. In the illustration shown, the signaling lipid phosphatidylinositol-3,4,5-trisphosphate (PIP_3) is synthesized specifically at the leading edge of a neutrophil migrating in the direction shown in the left panel (red arrow), while PIP_3 is degraded in other regions of the cell, creating the gradient of PIP_3 shown (in pseudocolor) in the right panel (reproduced with permission from Nishioka et al., *Mol. Biol. Cell* 19 [2008], 19, 4213–4223).

dissipate through lateral diffusion of MHC-1 molecules from the site of fusion [76, 77]. The dimensions of the spatial gradient generated by this mechanism will depend on the diffusion coefficient of the protein, which as already noted may be influenced by interactions with the underlying membrane cytoskeleton. A recent paper [78] has proposed a second mechanism by which submembrane cytoskeletal elements could influence the membrane distribution of some plasma membrane proteins. In this model, membrane proteins that interact (directly or indirectly) with actin microfilaments could be continuously transported to the cores of actin-based 'asters' at the plasma membrane, leading to transient clustering of such membrane proteins at the 'aster' core.

The examples discussed above illustrate the importance of the fact that biological membranes are subject to constant turnover and remodeling of their content, through both *de novo* creation or interconversions of components and transfer of materials between membranes through vesicular and non-vesicular trafficking pathways. At the same time, they illustrate how mechanisms of 'domain' formation based on *non-equilibrium* phenomena, which are often very difficult to study using model systems, can now be elucidated using increasingly sophisticated and quantitative experimental tools to study the behavior of membrane components in living cells.

·   ·   ·   ·

CHAPTER 4

Historical Development of the Concept of Membrane Nanodomains

By the early 1970s evidence that membranes are highly dynamic structures, allowing rapid lateral diffusion of various membrane components, led to the 'fluid mosaic' model of Singer and Nicolson [79], which has since become the dominant paradigm for envisioning biomembrane structure. The (rightful) emphasis of this model on the dynamic character of membranes was sometimes interpreted, overly simplistically, to imply that most membrane components intermix more or less randomly within the membrane plane. By the mid-1980s, this view was particularly widespread for the lipid components of membranes, which at physiological temperatures were considered to mix homogeneously in what would now be called a fluid-disordered bilayer phase. Speculation about the lateral organization of membrane proteins, particularly on the spatial 'mesoscale' (ca. 10–250 nm), was more discreet, in part because very few experimental tools were available to examine this issue for specific proteins. However, during the 1980s electron-microscopic studies clearly established the existence of various types of structurally well-defined, relatively long-lived 'domains' in membranes, including coated pits, caveolae and focal adhesions.

Renewed speculation that lipids might not always mix homogeneously in membranes arose from biophysical studies demonstrating that membrane sphingolipids differ strongly from typical membrane phospholipids in their physical properties (see Section 2.1). Thompson and Tillack discussed this point and its potential implications in an influential review published in 1985 [2], in which they also summarized early evidence suggesting that glycosphingolipids might be inhomogeneously distributed within cellular membranes. However, at the time potential lipid segregation in membranes was still envisioned mainly in terms of fluid–solid phase separations, a possibility considered highly unlikely for mammalian cell membranes. This conceptual difficulty was resolved a few years later with the publication of theoretical and experimental studies that demonstrated the formation of liquid-ordered phases, and the potential for segregation of liquid-ordered from liquid-disordered lipid phases, in sterol-containing lipid bilayers [80,81].

The first context in which segregation of liquid-ordered and liquid-disordered membrane domains was suggested to be functionally important was in the selective trafficking of glycosphingolipids from the Golgi to the apical membrane in polarized epithelial cells. Van Meer and Simons

[82, 83] proposed that within the Golgi complex glycosphingolipids cluster into patches that are selectively packaged into apically targeted transport vesicles; they also suggested that some apically targeted membrane proteins might be sorted into the same transport vesicles through association with the glycosphingolipid clusters. Consistent with these proposals, in 1992 Brown and Rose [84] reported that one class of apically targeted proteins, glycosylphosphatidylinositol-anchored proteins (GPI-proteins), could be isolated together with glycosphingolipids in a cold detergent-insoluble, low-density fraction obtained from these cells. Newly synthesized GPI-proteins could be isolated in this 'detergent-resistant membrane' (DRM) fraction only after they were transported from the endoplasmic reticulum to the Golgi, the site of glycosphingolipid synthesis. In subsequent work, Brown and colleagues also showed that GPI-proteins reconstituted into lipid vesicles could be isolated in a DRM fraction when, and only when, the vesicles incorporated liquid-ordered lipids [85]. DRM fractions isolated from cell membranes were enriched in lipids expected to be present in liquid-ordered domains (sterol and sphingolipids), and in some membrane proteins such as GPI-proteins and S-acylated proteins, but were strongly depleted in many other proteins, including many (but not all) transmembrane proteins [85, 86].

Due to its simplicity, the technique of isolating DRM fractions from cells rapidly became a popular means to identify cell membrane components that might be associated with liquid-ordered nanodomains in cell membranes under physiological conditions. By the mid-1990s, a rapidly growing literature reported many examples of proteins that were enriched in DRM fractions obtained from various cells and that were accordingly suggested to be associated with liquid-ordered lipid domains, by then termed 'lipid rafts,' in native cell membranes. However, many researchers recognized that DRM fractions could not automatically be equated with liquid-ordered domains hypothetically present in biological membranes [87], for reasons discussed in Section 5.1. Accordingly, other techniques, including various microscopic and spectroscopic methods, were explored to assess independently the mesoscale organization of membranes, in the process revealing evidence for new types of membrane nanodomains, such as the 'tetraspanin web' [53]. Isolation of DRM fractions nonetheless became a very popular means to assess the possible association of membrane components with rafts, in some cases without obtaining adequate corroborative evidence using other techniques.

By 2000, the raft concept had attracted great interest among cell biologists, and rafts were proposed, based on various lines of evidence, to play important roles in a variety of membrane-associated signaling processes, notably in cells of the immune system, and membrane-trafficking phenomena [88–90]. This excitement was reinforced by reports in which fluorescence microscopy was used to demonstrate segregation of micron-sized domains of liquid-ordered and liquid-disordered lipids in model membranes that mimicked, in a simplified manner, the lipid composition of the plasma membrane. As discussed in Section 2.1, biophysical studies of lipid and lipid–protein model systems have since revealed many interesting aspects of the properties of liquid-ordered lipid phases, at least some of which appear germane to understanding raft phenomena in biological membranes. However, the past decade has also seen considerable debate regarding the functional importance, the characteristics and even the existence of 'rafts' in various biological contexts [91–95]. At the same time, novel methodologies have provided important new insights into the diversity and functions of membrane nanodomains, including their generation through various combinations of molecular interactions (lipid–lipid, protein–lipid, protein–protein and even protein–carbohydrate), in a variety of biological contexts.

●   ●   ●   ●

CHAPTER 5

Experimental Tools to Study Nanodomain Formation in Cells

The methods described in the following sections are presented roughly in order of their historical introduction to the study of membrane nanodomains. As we will see, many of these techniques were first applied to study membrane 'rafts,' although some have proven very useful to characterize other types of membrane nanodomains as well.

5.1   BIOCHEMICAL FRACTIONATION METHODS

Treatment of animal or yeast cells (or isolated membranes) with certain nonionic detergents in the cold solubilizes a large proportion of membrane lipids but leaves an unsolubilized detergent-resistant membrane (DRM) fraction that can be isolated as a low-density fraction by gradient centrifugation. As noted in the last section, these 'detergent-resistant membrane' (DRM) fractions were initially suggested to arise from membrane rafts. The most commonly used method for isolation of 'raft' fractions uses the nonionic detergent Triton X-100, but a variety of alternative protocols using other detergents, or no detergent, have been reported. 'Raft' fractions obtained using different detergents, different concentrations of the same detergent (giving different ratios of detergent to membrane lipids), or different detergent-free methods often differ substantially in composition [86, 96].

The technical simplicity of detergent-based methods for isolating 'raft' fractions, and data from model systems suggesting that DRM fractions can be isolated only from membranes containing liquid-ordered domains [85], have made such methods very popular to assess the association of different membrane molecules with rafts and often, at least implicitly, to assess the presence of rafts in different cells and membranes. However, a variety of evidence suggests that biochemically isolated 'raft' fractions are not directly representative of rafts as they exist in biological membranes. First, isolated 'raft' fractions typically consist of membrane fragments with sizes approaching micron dimensions, while for most cell types microscopy does not provide evidence for raft domains anywhere near this size. Lower temperatures strongly enhance the formation and the size of liquid-ordered domains in lipid model systems [97, 98], and there is evidence for similar behavior in biological membranes [99, 100]. Detergents likewise cause enlargement, if not *de novo* creation, of liquid-ordered domains in lipid model membranes [101, 102]. The conditions used for most preparations of 'raft' fractions may therefore promote formation of segregated membrane lipid domains far larger than any present under physiological conditions. Second, it is questionable both on theoretical and on empirical grounds whether detergent treatments will ensure that all *bona fide* raft components remain associated with isolated DRMs, while all non-raft membrane components become solubilized, with high efficiency and fidelity. Some membrane components, for example, may be associated with rafts through interactions that are particularly susceptible to disruption by detergents [103]. Findings that DRM fractions isolated using different detergents or protocols

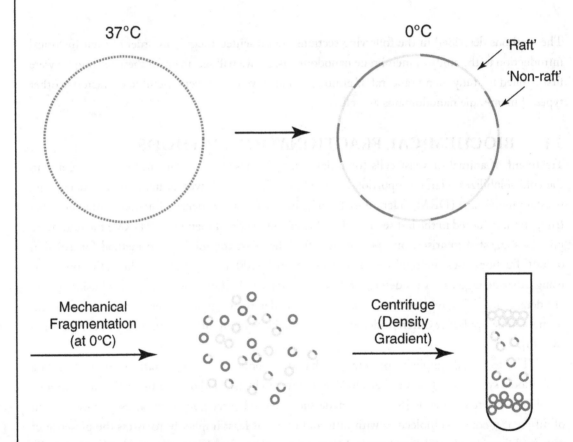

FIGURE 14: The physical bases for detergent-free preparations of 'raft' membrane fractions remain to be fully defined. A plausible, but hypothetical mechanism is illustrated in this figure. At low temperatures, membranes may form very large raft and non-raft domains. When the membrane is then mechanically disrupted to form smaller vesicles, some vesicles may contain mainly 'raft' components and others mainly 'non-raft' components. Also as shown, a substantial fraction of the vesicles is expected to contain mixtures of 'raft' and 'non-raft' components. Subsequent density-gradient centrifugation may provide a membrane fraction enriched in 'raft' components.

often differ substantially in composition further underscore potential concerns as to how faithfully 'raft' preparations obtained using detergents reflect the composition of rafts *in situ.*

While detergent-free methods may appear to be a preferable alternative to detergent-based methods to isolate 'raft' fractions, the changes in membrane organization that occur during preparation of detergent-free 'raft' fractions are also poorly understood. Detergent-free 'raft' preparations usually employ a combination of low temperature (0–4°C) and mechanical shearing to convert membranes into small vesicles. It is possible that cold treatment increases the size of liquid-ordered regions in membranes to such a degree that some of the vesicles produced upon membrane disruption could be largely or entirely derived from these regions of the membrane (Figure 14). However, the validity of this concept, or any other model for the formation of isolable 'raft' fractions using detergent-free procedures, remains to be assessed experimentally.

For all the reasons noted above, biochemical isolation of 'raft' fractions is now considered to be at best a somewhat unreliable approach to assess either the existence or the composition of rafts in a given membrane or cell type, notwithstanding the important role that this approach played in the historical development of the field. At a minimum other, more rigorous biophysical or imaging approaches like those discussed in the following subsections must be used to corroborate any conclusions suggested by analysis of biochemically isolated 'raft' fractions. This is often but not always the case in the current literature.

5.2 FLUORESCENCE MICROSCOPY AND COLOCALIZATION ANALYSIS

Most early studies of membrane nanodomains using conventional fluorescence microscopy focused on investigations of rafts, particularly in the plasma membrane, as discussed in this section. More recently developed fluorescence-microscopic techniques, using single-molecule or super-resolution fluorescence microscopy, have been used to examine formation of various types of membrane nanodomains, not all related to rafts, as discussed in Sections 5.8 and 5.9.

Early fluorescence-microscopic studies demonstrated that putative raft markers, such as GPI-anchored proteins, typically showed homogeneous distributions in the plasma membrane. Such evidence indicated that rafts, if present in the plasma membrane, must be smaller than the resolution limit of conventional optical microscopy (roughly 200–250 nm) and must be distributed fairly homogeneously within the membrane. Subsequent studies used antibodies or other multivalent proteins to crosslink ('patch') putative raft components on the cell surface, seeking thereby to promote aggregation of rafts into larger, inhomogeneously distributed structures that could be visualized by microscopy [61, 104]. Consistent with a model of small rafts coalescing into larger structures when raft components were crosslinked, it was found that independently crosslinking two putative raft components on the surface of fibroblasts or T-lymphocytes led to their colocalization in patches, while independently crosslinking a raft and a non-raft component did not [104].

Interestingly, crosslinking the plasma membrane GPI-protein placental alkaline phosphatase (PLAP), a putative raft marker anchored only to the outer leaflet of the membrane, induced formation of PLAP-enriched patches that were also enriched in the acylated protein tyrosine kinase Fyn, a raft marker anchored only to the inner membrane leaflet. Such observations suggested that rafts could be membrane-spanning structures.

The results of co-patching experiments like those just described are intriguing but may not prove that rafts are present in cell membranes under 'basal' conditions. As noted in an earlier section, multivalent binding of proteins to membrane components (e.g., of the pentavalent cholera toxin B-subunit to ganglioside GM1) under some conditions can trigger segregation of lipid domains that are not observed in the absence of the protein [12]. It is also known that crosslinking GPI-anchored proteins on the cell surface can trigger interactions with the actin cytoskeleton on the other side of the membrane and ultimately induce calcium signaling within the cell [105]. It is therefore possible that 'patching' membrane components may promote reorganization or even *de novo* formation, rather than simple lateral aggregation, of membrane domains. For these reasons, studies of this type now typically use treatments that promote only very small-scale clustering of membrane components [105–107].

A further complication in interpreting apparent colocalization of different membrane components, whether 'patching' treatments are used or not, is the fact that local variations in the topography of a membrane (local 'wrinkling' of the membrane, projections, such as microvilli, etc.) can create significant spatial variations in the amount of membrane imaged in different pixels (or voxels), as illustrated in Figure 15. This can lead to apparent spatial variations in the distribution of membrane components that actually reflect local variations in the morphology or orientation of the membrane itself [108, 110]. One means to correct for such effects is to compare the spatial distribution of a membrane component of interest to that of a fluorescent 'reference' marker that distributes homogeneously within the membrane. Fluorescent species that interact nonspecifically with membrane lipids can be used as reference markers for this purpose [108].

Another interesting fluorescence-based approach to investigate membrane heterogeneity in living cells uses lipophilic fluorescent probes whose fluorescence properties are sensitive to the local physical state of the membrane lipids. The most widely used probes of this type are Laurdan and structurally related species, which can be incorporated into cellular membranes when added exogenously. Laurdan provides information about the polarity (water content and dynamics) and order of its immediate lipid environment through measurements of a fluorescent parameter known as the 'generalized polarization,' determined by measuring the ratio of fluorescence emitted over different ranges of wavelengths. This probe has been used, among other applications, to provide evidence that the average ordering of the membrane is enhanced, consistent with an accumulation of rafts, in regions of contact between T-lymphocytes and antigen-coated beads [110].

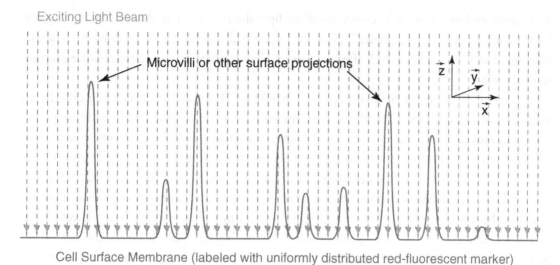

Fluorescent Image (linescan shown on black background)

FIGURE 15: Topographical features of membranes affect fluorescence imaging. A scanning excitation beam is shown illuminating a narrow strip of membrane whose cross-sectional topographical profile is shown in red. In regions where the membrane surface is tilted closer to the direction of the exciting light (**z**), the exciting beam 'sees' a greater number of membrane-bound fluorescent molecules per unit area in the *xy*-plane, and the fluorescence recorded from that region will be correspondingly greater, as illustrated by the expected linescan shown. Since many features of membrane topography are smaller than the resolution limits of fluorescence microscopes, the topographical origins of these effects may not be obvious, and from an image like that shown it might be concluded that a fluorescent marker is concentrated in particular membrane 'domains,' when in fact it is homogeneously distributed in the membrane.

5.3   MANIPULATIONS OF MEMBRANE LIPID COMPOSITION

Perturbations of membrane lipid composition have been used as an experimental tool in studies of membrane rafts, which as noted earlier are postulated to arise in part through lipid–lipid interactions. Since formation of lipid-ordered domains in model membranes requires the simultaneous presence of both 'high-melting' lipids and cholesterol (or a related sterol), perturbing the levels of sphingolipids or sterol in a membrane would be expected to alter the extent of lipid domain segregation and/or the physical properties of the lipid domains present. Researchers have accordingly examined how various physical, pharmacological and metabolic perturbations of membrane cholesterol and sphingolipid levels affect diverse cellular processes in which lipid rafts are proposed to play a role.

Various methods can be used to perturb membrane cholesterol levels in living cells. In one very popular method, cells are incubated for relatively short times (typically <1 h) with compounds known as cyclodextrins, which comprise 6–8 monosaccharide residues joined to form a ring-like cyclic oligomer. Cyclodextrins have a hydrophobic internal cavity that can bind cholesterol and related sterols; derivatives of the heptameric β-cyclodextrin are most commonly used for this purpose. When incubated with cyclodextrins, cells rapidly lose a significant fraction of their total cholesterol, which is extracted into the extracellular medium to form soluble cholesterol/cyclodextrin complexes. Cholesterol transfers rather rapidly between different cellular membranes via various processes [111], and cholesterol extracted from cells by cyclodextrins may therefore originate not only from the plasma membrane but also from intracellular compartments. Incubation of cells with sterol/cyclodextrin complexes allows partial replacement of cellular cholesterol with the exogenous sterol, providing a useful method to incorporate cholesterol analogues into cells [112].

Cyclodextrin treatment has been used in many studies to reduce cellular cholesterol levels and has been reported to cause significant perturbations of a number of cellular and membrane functions. However, for various reasons cyclodextrin treatment cannot be considered a highly selective method to perturb raft-dependent phenomena. First, cholesterol depletion has been shown to perturb various membrane functions that have no evident connection to raft-dependent phenomena, such as clathrin-dependent endocytosis [113]. Second, cyclodextrin treatment has been found to alter some cellular properties that play broad roles in cellular regulation (e.g., the integrity of intracellular calcium stores and the plasma membrane potential in T-lymphocytes [114]); perturbations of such central elements in cellular function could perturb a variety of 'downstream' processes that are not themselves raft- or even cholesterol-dependent. Third, cyclodextrins have been shown to alter at least one membrane property, the rates of lateral diffusion of proteins on the cell surface, in a manner that is independent of their cholesterol-binding ability [115]. All of these points illustrate that cyclodextrins may in principle alter a given membrane function through mechanisms unrelated to perturbation of rafts or even, in some cases, unrelated to reduction of cellular cholesterol levels. Some, but not all potential concerns attending the use of cyclodextrins may be assessed through control incubations of cells with mixtures of cyclodextrins and cholesterol in a stoichiometry that does not cause net depletion of cellular cholesterol. A better, but infrequently used experimental approach is to compare how a phenomenon of interest is affected by incubating cells with complexes of cyclodextrins with different cholesterol analogues that support or inhibit formation of liquid-ordered domains [116].

An alternative method to alter cellular cholesterol levels is to culture cells in lipoprotein-deficient media in the presence of inhibitors of cholesterol biosynthesis. This approach provides a valuable alternative to cyclodextrin treatment to deplete cellular cholesterol. However, since such 'metabolic' treatments to reduce cellular cholesterol typically extend over periods of days, cells may be able to adapt more extensively (for better or for worse, depending on the precise experimental objective) to depletion of sterol levels than they can in response to shorter-term treatment with cyclodextrins. In experiments designed to assess the cholesterol dependence of a given cellular process, it is therefore typically desirable to compare the effects of cyclodextrin treatment vs. metabolic approaches to deplete cellular cholesterol levels. It is also important to note, of course, that the finding that a given phenomenon is dependent upon cholesterol (or any particular lipid or group of lipids) does not automatically imply that it is raft-dependent.

Biosynthesis of various membrane sphingolipids can be inhibited both by pharmacological and by genetic approaches. Broad inhibition of sphingolipid synthesis, using either PDMP (DL-threo-1-phenyl-2-decanoylamino-3-morpholino-1-propanol), an inhibitor of glycosylceramide synthase, or fumonisin B1, an inhibitor of ceramide synthases, has been used to obtain evidence that sphingolipids play important roles in a variety of membrane functions [117]. Gene deletions in mice have been reported that abolish synthesis of several different groups of glycosphingolipids, including galactocerebrosides, sulfatides and various families of gangliosides [118]. Deficiencies in individual classes of these glycosphingolipids typically lead to severe neurological and/or reproductive defects, while homozygous deletion of the glucosylceramide synthase gene, which abolishes synthesis of almost all glycosphingolipids, leads to embryonic lethality. Mice and mouse cell lines deficient in various glycosphingolipids provide useful tools to study the biological functions of these lipids, potentially including their contributions to raft-dependent phenomena. Interestingly, a recent genome-wide siRNA screen in a simpler organism, the nematode *C. elegans,* identified several enzymes involved in glycosphingolipid synthesis as essential for normal differentiation of apical from basolateral membranes in epithelial cells of the gut, and consequently for development of normal gut morphology [119].

Yeast, and particularly the familiar budding yeast *Saccharomyces cerevisiae,* has been widely used in both targeted and genome-wide studies that have identified important roles for particular classes of lipids in phenomena such as membrane trafficking. Such studies have for example identified ergosterol as well as lipids containing very long-chain (C26) fatty acids as important factors for proper sorting and trafficking of particular integral membrane proteins through the secretory pathway to the cell surface [120, 121].

5.4   LIPID MODEL SYSTEMS

As discussed in Section 2.1, three-component bilayers formed from mixtures of cholesterol, a saturated phospho- or sphingolipid and an unsaturated phospholipid exhibit coexisting liquid-ordered (Lo) and liquid-disordered (Ld) domains, which at subphysiological temperatures (typically <25°C) are large enough to be observed by conventional fluorescence microscopy (i.e., >ca. 200 nm)

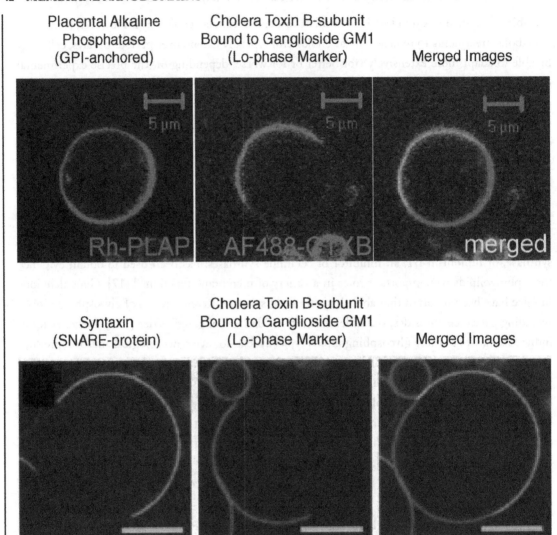

FIGURE 16: Partitioning of membrane proteins between liquid-ordered and liquid-disordered lipid domains can be visualized by fluorescence microscopy of giant lipid/protein vesicles. The confocal-microscopic images shown represent cross-sections through giant unilamellar vesicles that exhibit Ld/Lo phase separation and that incorporate either the GPI-anchored protein placental alkaline phosphatase (PLAP, top panel) or syntaxin, a transmembrane SNARE protein involved in membrane fusion (bottom panel). The vesicles are also labeled with fluorescent cholera toxin B-subunit (CTxB), bound to ganglioside GM1, as a selective marker for Lo-phase domains (middle image in each panel). PLAP partitions partly into Lo-domains labeled by CTxB and partly into Ld domains from which CTxB is absent. By contrast, syntaxin is strongly excluded from the Lo domains (top and bottom panels reproduced with permission from references [30] and [122], respectively).

in giant unilamellar vesicles. When fluorescent-labeled proteins are incorporated into such vesicles, their distributions between the Lo and Ld domains can be observed, as illustrated in Figure 16. The GPI-anchored placental alkaline phosphatase, for example, was shown by this method to partition significantly, though only partially, into Lo domains in vesicles containing both Lo and Ld domains, while in similar vesicles the transmembrane fusion (SNARE) proteins syntaxin and synaptobrevin showed a strong preference for liquid-disordered domains [32, 122]. For technical reasons this approach cannot determine quantitatively the relative affinity of a given membrane component for the Lo vs. the Ld phase, but is a useful tool to assess this property qualitatively.

An interesting variation of the above approach, already described in Section 2.2.4, is to measure by microscopy the distribution of fluorescent-labeled lipids (and, potentially, proteins) between giant unilamellar vesicles and narrow, connected tubules ('tethers') pulled from these vesicles [44–46]. Under appropriate conditions, the curvature of the 'tether' can be adjusted, to span a range of values comparable to that found for tubular membrane structures within cells, by adjusting the balance of mechanical forces in the system. This methodology offers a very useful approach to assess the distributions of membrane components between membrane regions of different curvatures, an issue important for understanding the formation and properties of various curved membrane 'domains' within the cell.

Fluorescence-spectroscopic approaches have also been used to compare the partitioning of different membrane molecules between Lo and Ld domains in model membranes [37, 123]. These techniques again examine the distributions of fluorescent-labeled membrane molecules in lipid model membranes that exhibit coexisting Lo and Ld domains. However, in this case, the distribution of the fluorescent species between domains is monitored not by direct microscopic observation but rather through differences in the fluorescence signal from the labeled species when present in the Lo vs. the Ld phase. This approach has been used to compare the affinities of a variety of lipids and peptides for Lo vs. Ld domains [36, 37, 123, 124].

5.5   ELECTRON-MICROSCOPIC METHODS AND ANALYSIS OF MOLECULAR CLUSTERING

Electron microscopy (EM), which offers spatial resolution much greater than that of conventional light microscopy, has been used to examine and to compare the distributions of putative raft and nonraft components in the plasma membrane of various types of cells. While EM cannot visualize individual biomolecules directly in complex systems like cells, it can readily visualize small gold particles, which can be attached to antibodies and thereby used as markers for specific proteins or lipids. Different proteins or lipids can moreover be labeled in the same sample using multiple antibodies coupled to gold particles of different sizes (e.g., 6 nm and 12 nm diameter), which can be readily

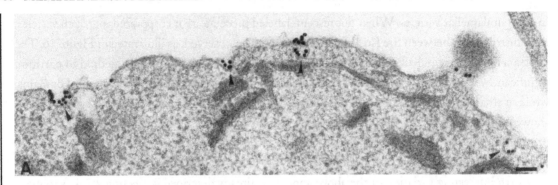

FIGURE 17: Gold-nanoparticle labeling and electron microscopy allow localization of membrane proteins with high spatial resolution. The image shows a small section of the plasma membrane of A431 cells labeled with cholera toxin conjugated to 14-nm colloidal gold particles, seen as black dots. The particles, which bind to ganglioside GM1 in the membrane, are highly concentrated in membrane invaginations that were identified as caveolae through further experiments (reproduced with permission from Parton, *J. Histochem. Cytochem.* 42 [1994], 155–166). Space bar = 100 nm.

distinguished in EM images. Early studies of plasma membrane organization used this immunogold-labeling methodology mainly to map the distributions of particular membrane proteins and lipids on cell surfaces in relationship to morphologically identifiable membrane structures such as coated pits and caveolae [125]. Such studies were particularly useful to examine the association of various putative raft components with caveolae (Figure 17), which revealed similarities but also noteworthy differences between the compositions of caveolae and rafts (see the book by Stan in this series). One intriguing study of this type found that antibody-induced clustering of GPI-anchored membrane proteins could induce their translocation to caveolae [61].

FIGURE 18: Immunogold-electron microscopy and the 'rip-off' technique allow high-resolution imaging of protein distributions in the plasma membrane. (A) Cells are 'sandwiched' between a coverslip and an EM grid, adhering tightly to both. The coverslip and grid are then pulled apart, disrupting the cells and leaving large plasma membrane fragments adsorbed to the grid. Specific proteins (or lipids) exposed at the cytoplasmic face of these membrane fragments can then be labeled with antibody–gold complexes, whose distributions in the membrane plane are subsequently imaged by electron microscopy as shown. Alternatively, proteins (or lipids) present at the outer surface of the plasma membrane can be imaged by immunogold-labeling cells before disrupting them (not shown). (B) Electron micrograph showing the plasma membrane distribution of GFP fused to the membrane-anchoring sequence of H-ras (reproduced with permission from reference [126]). Space bar = 100 nm.

A

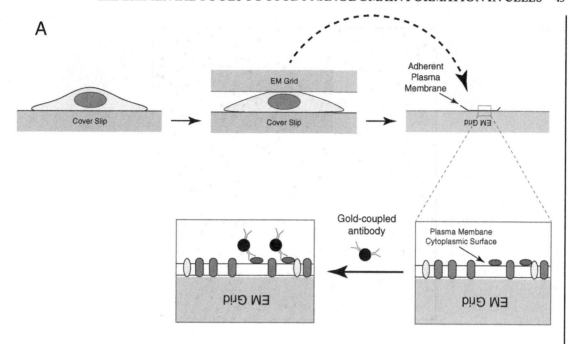

B

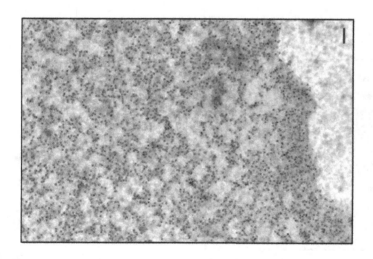

A

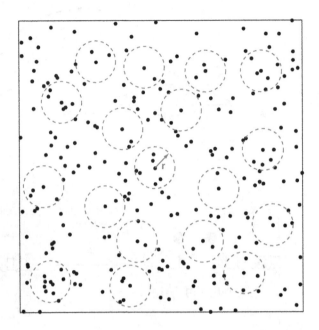

B

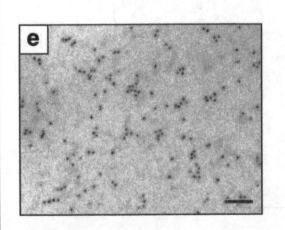

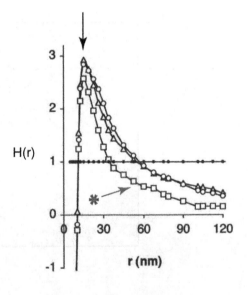

Most of the studies noted above used classical thin-sectioning methods for sample preparation and consequently imaged membranes mainly in cross- or oblique sections. More powerful EM-based approaches for imaging the lateral distributions of molecules within membranes have emerged with the development of 'rip-off' methods that afford *en face* views of extensive membrane regions labeled with antibody-gold markers for specific membrane components. In these methods, cells adhering to a coverslip are also induced to adhere to a suitably treated EM grid, which is then pulled away from the coverslip, leaving large membrane fragments adherent to the EM grid with their cytoplasmic surfaces directed outward (Figure 18A). Membrane components present at the extracytoplasmic or the cytoplasmic faces of the plasma membrane can be labeled by incubation with antibody-gold complexes before or after, respectively, the grid-adhesion/shearing stage of the procedure. An example of such labeling, shown for the cytofacially associated protein H-*ras* [126], is shown in Figure 18B.

En face images revealing the lateral distributions of specific proteins (or lipids) in membranes constitute very rich sources of information, only a small portion of which is apparent through simple visual inspection. Such images yield much more information when analyzed using computational methods to assess quantitatively the clustering (or dispersion) of a single membrane component, or of the distributions of two different components with respect to one another. We will consider one such method below, as it constitutes an excellent example of how quantitative analyses are expanding the power of microscopic approaches to address important questions in cell biology.

The essential concepts underlying quantitative analysis of particle distributions by the method we will consider here, based on derivatives of a function known as Ripley's K-statistic, can be understood with reference to Figure 19A, which represents schematically a flat area of membrane

FIGURE 19: Quantitative analyses of the distributions of membrane proteins or lipids can provide information that is not apparent from visual examination. (A) Schematic illustrating the basis on which Ripley's K-statistic and related quantities are determined as discussed in the text. (B) Left: immunogold-EM image showing the distribution of GFP-tK, a construct fusing GFP to the membrane-binding sequence of K-ras, at the cytoplasmic surface of a cell plasma membrane. Right: Plots of $H(r) = (\hat{E}(r) / \lambda\pi)^{1/2} - r$ (see text for description) for this GFP-tK construct in cultured cells. The curve marked with a red asterisk (open squares) was obtained for cells treated for 1 h with cyclodextrin to deplete plasma membrane cholesterol, while the top curve (open circles) was obtained for untreated cells. Cyclodextrin treatment induces a significant decrease in the extent of clustering of GFP-tK (indicated by the decrease in $H(r)$ over a broad range of distances r) but does not alter the most probable distance between clustered GFP-tK molecules (indicated by the peak of each curve) (reproduced with permission from reference [126]).

on which a number of labeled molecules are distributed. (Note that the particles shown may seem to tend to cluster, but their distribution is in fact completely random—a simple illustration of the limitations of visual analysis to interpret such images.) As the heart of our analysis, we determine from an image (or set of images) a function represented as $\hat{E}(r)$, which represents the average or 'expected' number of particles that lie within a distance r of another particle in the image field. As illustrated in Figure 19A, if we create a set of circles of radius r centered on *every* particle in the image (only some of which are shown for clarity in the figure), $\hat{E}(r)$ will represent the *average* number of particles found in each circle (excluding the central particle in each case). This quantity is readily determined, as a function of r, by computational analysis of the experimental images. It is easy to show that if the distribution of particles is perfectly random, the value of $\hat{E}(r)$ is simply $\lambda \pi r^2$, where λ is the average density of labeled molecules in the image field, and that in this case the function $H(r) = (\hat{E}(r)/\lambda\pi)^{1/2} - r$ has a value of zero for all values of r. Experimentally determined plots of $H(r)$ (sometimes alternatively denoted as $L(r)$ - r) *vs.* r can readily be interpreted visually: if the value of $H(r)$ is significantly greater (/smaller) than zero over a given range of r-values, we can conclude that the particles exhibit a greater (/smaller) than random tendency to lie within this range of distances with respect to one another.

Figure 19B shows examples of experimentally determined plots of $H(r)$ vs. r, obtained by analysis of immunogold-labeling EM images for a membrane-associated protein, along with a representative EM image used in the analysis. All three plots shown reveal a significant tendency for the protein to self-cluster, as indicated by values of $H(r)$ that markedly exceed those expected for a random distribution of the protein (i.e., that lie well above the horizontal line shown in the experimental plots) over a range of separations centered at roughly 16 nm. Plots such as these are typically interpreted semiquantitatively (e.g., "the particles show a significant tendency to cluster over a range of distances from x to y") or in a comparative manner. More detailed analysis of such data requires curve fitting in the context of specific (but often somewhat abstract) models for the forces, constraints, etc. that govern the particle distribution. Such analyses can be useful to demonstrate that a particular model does not describe satisfactorily a given set of experimental data. However, a finding that the predictions of a given model describe a given set of data well does not prove that the model is the only one consistent with the data.

Analysis like those discussed above can easily be adapted to investigate the spatial relationships between molecules of different types within a membrane, to determine for example whether two species **X** and **Y** tend to concentrate together in clusters or, conversely, tend to segregate in distinct areas of the membrane. The overall principle of this analysis is similar to that described above to characterize the spatial relationships between molecules of a single type. In this case, however, the analysis is based on analyzing (again, as a function of r) the average or expected number of molecules of species **Y** that lie within a distance r of a molecule of species **X**, and *vice versa*.

Analyses of molecular distributions like those described above were first utilized to examine data from immunogold-EM experiments, which offer a degree of spatial resolution that was until recently unattainable by light microscopy. Newer fluorescence-microscopic approaches, such as the 'super-resolution' microscopies described in Section 5.9, now also can provide information about molecular distributions on distance scales as small as tens of nanometers, and data derived from fluorescence microscopy are therefore increasingly being subjected to quantitative analysis as well. Other types of statistical analyses have also been used to evaluate particle distributions and molecular clustering in membranes, including determination of pair correlation functions [127] and various forms of image correlation spectroscopy [128].

5.6   FLUORESCENCE RESONANCE ENERGY TRANSFER

As already noted, conventional fluorescence microscopy cannot resolve distances smaller than 200–250 nm and therefore cannot monitor interactions between molecules on distance scales closer to molecular dimensions (e.g., formation of oligomers or small clusters). However, fluorescent molecules that approach within roughly 10 nm or closer can exhibit a phenomenon known as fluorescence resonance energy transfer (often abbreviated to FRET or RET), which is frequently exploited to examine molecular interactions on such small distance scales. The principle of FRET is illustrated schematically in Figure 20. If a fluorescent 'donor' residue (D) is present in isolation, when it absorbs (i.e., is 'excited' by) photons with a characteristic excitation wavelength $\lambda_{exc}(D)$, it will emit photons, i.e., fluoresce, at a characteristic longer emission wavelength $\lambda_{em}(D)$. However, if an appropriate 'acceptor' residue A lies sufficiently close to the residue of D (we will shortly define the terms 'appropriate' and 'sufficiently'), through resonance energy transfer the acceptor residue A can 'siphon off' a fraction of the energy absorbed (as photons) by D, thereby reducing the number of photons that D emits as fluorescence. Moreover, if A itself is a fluorescent species, it may use the energy 'siphoned off' from D to emit fluorescence at its own characteristic emission wavelength $\lambda_{em}(A)$. FRET between donor and acceptor fluorescent groups can be monitored in multiple ways, e.g., through a decrease in the intensity of fluorescence emitted by D when A is present or, if A is also fluorescent, through the intensity of fluorescence emitted by A when photons are absorbed by D.

In order for two residues D and A to exhibit FRET, two important conditions must be fulfilled. First, the emission (fluorescence) spectrum of D must overlap significantly with the absorption spectrum of A, as illustrated in Figure 20. Second, D and A must lie within a sufficiently small distance R (typically several nm or less). These conditions can be summarized more quantitatively by the following equation for a pair of randomly oriented donor and acceptor molecules separated by a distance R:

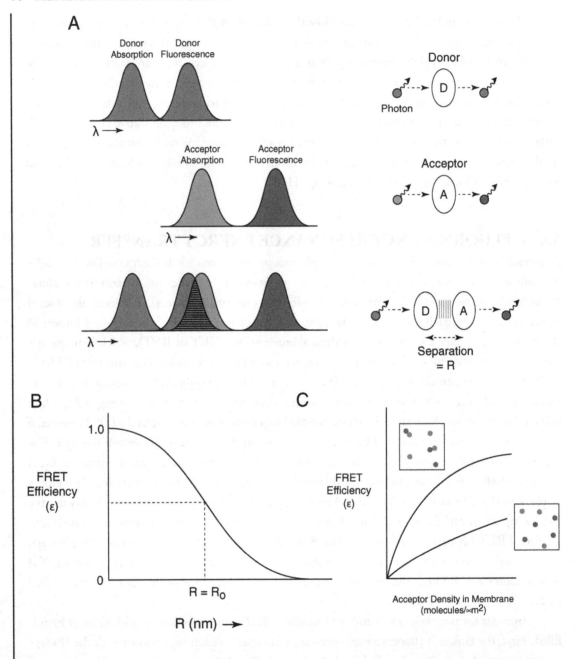

$$\text{FRET efficiency } (\varepsilon) = \frac{1}{1+(R/R_\text{o})^6}$$

Where ε represents the efficiency with which A 'siphons off' the energy of photons absorbed by D, R is the distance between D and A and R_o is a constant (for any given type of donor–acceptor pair) called the Förster length. The value of ε can range from 0 (no FRET) to 1 ('perfectly efficient' FRET, *i.e.*, the fluorescence of **D** is entirely suppressed by the presence of **A**). As illustrated in Figure 20B, the efficiency of energy transfer varies strongly with the donor–acceptor separation R over a range of distances extending up to roughly two to three times the value of R_o. R_o values for typical donor–acceptor pairs range up to ca. 6 nm, and FRET can accordingly be used to examine the spatial relationships between molecules on distance scales up to roughly 20 nm. Modifications of the above equation can be derived to describe the efficiency of FRET between populations of donor- and acceptor-labeled molecules distributed in various manners (randomly, as defined oligomers, enriched in small domains or clusters, etc.) within a membrane. Experimental FRET data can thus be compared to model predictions to assess, for example, whether a given membrane protein (or lipid) exists within the cell membrane as monomers or as oligomers.

FRET-based studies have sought both to assess whether 'rafts' are present in mammalian cell membranes and to estimate the dimensions of such domains if present. In a typical experiment of this type, one or two putative raft components (e.g., specific GPI-anchored proteins or gangliosides)

FIGURE 20: Fluorescence resonance energy transfer (FRET) can monitor the spatial relationships between different membrane molecules. (A) Different fluorescent molecules can serve as FRET donors and acceptors when the fluorescence emission spectrum of one species (donor) overlaps the excitation spectrum of the other (acceptor), as shown by the hatched area in the lower set of spectra. When a donor- and an acceptor-labeled molecule are in sufficiently close proximity, energy absorbed (as photons) by the donor can be transferred to the acceptor, quenching the donor fluorescence and allowing the acceptor to fluoresce when the donor is excited by illuminating light. (B) The efficiency of FRET between donor- and acceptor-labeled molecules varies strongly with the donor–acceptor distance, R, over a nanometer-scale range of distances. The quantitative nature of this relationship depends (through the parameter R_o) on the specific donor and acceptor used, as discussed in the text. (C) Clustering or other nonrandom distributions of membrane proteins or lipids can be detected by measuring FRET efficiency as a function of the density of the acceptor-labeled species in the membrane. The upper versus the lower curve shown could be observed if a pair of donor- and acceptor-labeled species either interact/cluster with each other (upper curve) or, by contrast, are randomly and independently distributed in the membrane (lower curve).

are labeled with appropriate donor- and acceptor-tagged proteins or ligands. Using fluorescence microscopy, the extent of FRET from the donor- to the acceptor-labeled species is then measured as a function of the surface density of the acceptor-labeled species (which varies naturally from cell to cell and can be further varied, if necessary, by adjusting the experimental conditions). The relationship between FRET efficiency and the surface density of the acceptor-labeled species can then be analyzed to assess whether the donor- and acceptor-labeled species are clustered or enriched in small domains or are randomly distributed in the membrane, as illustrated in Figure 20C. Different FRET studies of 'raft' markers have produced varying results, with some studies concluding that these species are randomly distributed in the plasma membrane while others have reported that they show significant co-clustering, consistent with enrichment in particular membrane nanodomains [129–133].

5.7 BIMOLECULAR FRAGMENT COMPLEMENTATION (BIFC) MICROSCOPY

Green fluorescent protein (GFP) and related fluorescent proteins can assemble spontaneously from two independently expressed fragments if the fragments are brought into close proximity. This effect can be used to detect interactions between monomers of a single protein or between two different proteins in living cells [134]. As illustrated in Figure 21, when two proteins fused to (nonfluorescent) complementary fragments of GFP are expressed in the same cell, association between the two proteins will allow assembly of the GFP fragments into 'complete' GFP modules whose fluorescence can be detected microscopically (Figure 21). BiFC provides a sensitive method that can detect even weak or transient protein–protein interactions and is therefore useful to detect oligomerization or clustering of proteins in membranes. Because interactions between the coupled 'reporter' GFP fragments can distort both the affinity and the kinetics of interactions between the protein species of interest, BiFC is typically used in a qualitative rather than a quantitative manner to demonstrate interactions between proteins in membranes. The method can nonetheless be a valuable means to demonstrate clustering or oligomerization of membrane proteins so long as these limitations are properly considered in experimental design and interpretation [135, 136].

5.8 SINGLE-PARTICLE MEASUREMENTS OF MOLECULAR DIFFUSION AND INTERACTIONS

When protein (or lipid) molecules on the cell surface are labeled at low surface densities with appropriate fluorophores or with small gold particles, the trajectories of individual labeled particles diffusing on the cell surface can be tracked by microscopy [137]. By analyzing such trajectories for a number of particles, it is possible to determine not only the average speed of diffusion of the particles on the surface but also whether particle diffusion on the surface is entirely free or

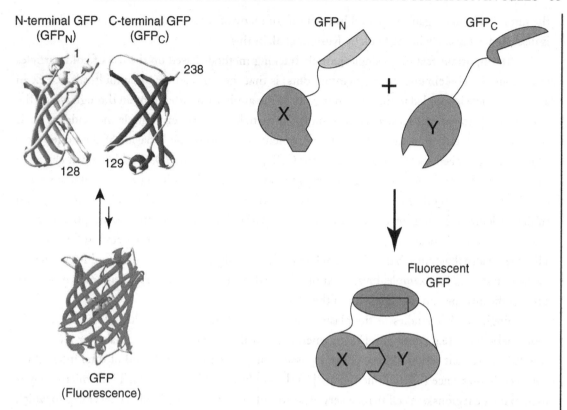

FIGURE 21: Bimolecular fragment complementation (BiFC) using 'split GFP' allows microscopic visualization of protein-protein interactions. As shown at the left, N- and C-terminal fragments of GFP (GFP$_N$ and GFP$_C$) can be expressed separately; each is nonfluorescent unless it reassembles with the complementary fragment. By themselves, GFP$_N$ and GFP$_C$ interact too weakly to produce a stable fluorescent protein when coexpressed in cells. However, as shown at the right, if two proteins **X** and **Y** associate within the cell, interaction between the **X**- and **Y**-portions of coexpressed **X**-GFP$_N$ and **Y**-GFP$_C$ constructs can promote and stabilize association of the GFP$_N$ and GFP$_C$ fragments to form a fluorescent GFP module (left side of figure reproduced with permission from Ozawa et al., *Analyt. Chim. Acta* 556 [2006], 58–68).

is constrained over particular scales of time or distance. Such information provides a much more detailed picture of the diffusion process, and indirectly, of the membrane milieu within which the particle is diffusing, than do population-based methods to monitor diffusion such as fluorescence recovery after photobleaching (FRAP) or fluorescence correlation spectroscopy (FCS). One of the key conclusions derived from single-particle studies is that the cytoskeleton underlying the plasma membrane influences the diffusion of many plasma membrane components, thereby modulating the intramembrane organization and interactions of many of these components as well. This topic is discussed extensively in the book by Kusumi et al. in this series.

An important feature of single-particle tracking methods based on the use of gold particles (and, under certain circumstances, quantum dots) is that the particles are often multivalent, i.e., an average particle is coupled to multiple molecules of the antibody or other protein through which the particle binds to the cell surface. As a result, such particles can cluster multiple molecules of their protein or lipid 'target' in the cell membrane, potentially altering the behavior of the target molecules. This has been shown to be the case for GPI-proteins such as CD59, which when clustered into very small aggregates in the plasma membrane can intermittently engage the submembrane cytoskeleton (apparently through interactions with transmembrane proteins), cease diffusing and initiate calcium signaling in the cell interior [105–107]. This is a very interesting phenomenon that can be physiologically important in some contexts, as discussed in more detail in Section 6.1. However, it also illustrates that multivalent ligands (potentially including not only multivalent nanoparticles but also, for example, bivalent antibodies) when used as probes for membrane molecules can significantly perturb the behavior of their 'targets.'

Single-particle studies of the plasma membrane diffusion of molecules labeled with small, monomeric 'tags' (e.g., antibody Fab fragments, fused fluorescent protein domains or chemically coupled fluorescent residues) have become possible only recently, with the aid of total internal reflectance fluorescence (TIRF) microscopy [138]. As illustrated in Figure 22, TIRF microscopy is used to image regions of a cell that lie very close to a glass surface (e.g., a slide or cover slip to which cells adhere), which is illuminated using laser light directed at a very shallow angle with respect to the surface. Under these conditions, the light does not pass through the glass-liquid interface but instead is reflected and creates in the liquid phase a so-called evanescent wave of light, whose intensity falls off exponentially with distance from the interface. The evanescent wave can excite fluorescent molecules found in the medium near the glass surface. Since the intensity of the wave falls off steeply with distance from the surface, fluorescent molecules located very close to the surface (typically <100 nm) are excited much more strongly, and hence are detected with much greater sensitivity, than molecules lying farther from the surface. TIRF microscopy therefore allows highly selective, and sensitive, detection of fluorescent molecules present in (or on) regions of the plasma membrane juxtaposed to the glass surface while minimizing interference (including autofluorescence) from other areas of the cell. This advantage is critical to allow detection of fluorescence from the plasma membrane at the single-molecule level.

Single-molecule observations of fluorescent-tagged molecules can provide information not only about molecular diffusion *per se* but also about intermolecular interactions in (or on) the plasma membrane, using either of two approaches (Figure 23). First, two fluorescent-labeled molecules tracked simultaneously can sometimes be observed to approach one another and to diffuse together

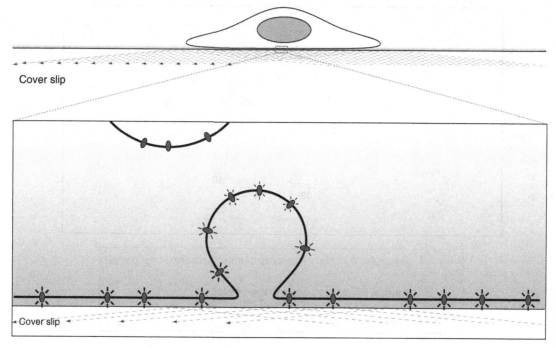

Cover slip

Cover slip

FIGURE 22: Total internal reflectance fluorescence (TIRF) microscopy selectively visualizes structures within or very near the plasma membrane. When a glass coverslip (shown with an adherent cell) is illuminated with light at a sufficiently low angle, the light beam reflects entirely off the glass/water surface rather than passing into the overlying aqueous phase. The reflected light beam creates an 'evanescent wave' (shown here as a gradient of green color), which decays very sharply with distance from the surface, at the glass/liquid interface. The evanescent wave can excite fluorescence from molecules lying within roughly 100 nm of the interface but not from molecules outside this region. TIRF can thus image fluorescent molecules in the plasma membrane where cells adhere to the coverslip ('halos' in the figure indicate fluorescing molecules), with minimal fluorescence interference from other regions within the cells.

(within the resolution limits of the method) for periods significantly longer than expected purely by chance (Figure 23A). With suitable statistical analysis, it is possible to assess from such results whether particular species interact within the membrane and to determine the kinetics (e.g., the average lifetime) of their interaction [135, 139]. Second, transient interactions between fluorescent particles can be detected by single-molecule FRET (see Section 5.6) if the two interacting species are labeled with suitable donor and acceptor groups. As shown in Figure 23B, when a donor-labeled molecule and an acceptor-labeled molecule associate on the membrane surface, a FRET signal will be detected and can be tracked (e.g., as emission of fluorescence at the acceptor's emission wavelength $\lambda_{em}(\mathbf{A})$ when the sample is illuminated at the donor's excitation wavelength $\lambda_{exc}(D)$). This

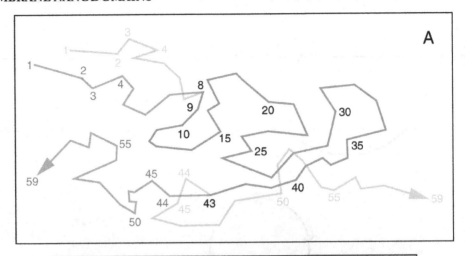

Particle A Trajectory (numbers denote successive time points)

Particle B Trajectory (same time points)

A and B Codiffusing

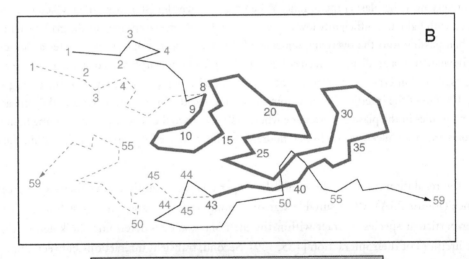

Particle A Trajectory

Particle B Trajectory

A and B Codiffusing (FRET Signal Observed)

FIGURE 23: Analysis of diffusion trajectories can reveal details of the interactions between membrane molecules. (A) Under appropriate conditions, the diffusion trajectories of individual fluorescent-labeled molecules in a membrane can be tracked by fluorescence microscopy. In the schematic example shown, the positions of two labeled molecules are imaged in parallel at a series of time points (from 1 to 59). Initially, the trajectories of the two molecules are separate (yellow, cyan), but from time points 8–43 they coincide as the molecules associate (green segment) before diverging again thereafter as the molecules dissociate. From such data, the thermodynamic and kinetic constants for association/dissociation of the molecules in the membrane can be determined. (B) Fluorescence resonance energy transfer (FRET) provides another means to monitor the interactions between individual molecules as they diffuse in the membrane. Two molecules, one labeled with a FRET donor and the other with a FRET acceptor, initially are diffusing separately (grey trajectories) and give no FRET signal. From time points 8–43, the molecules associate to form a dimer that produces a FRET signal (red segment) before the molecules dissociate and again diffuse separately.

approach is particularly powerful since FRET operates over short distance scales (< 20 nm) and thus can detect interactions between molecules in a fairly direct manner.

5.9   SUPER-RESOLUTION LIGHT MICROSCOPIES

As already noted, due to diffraction effects, conventional fluorescence microscopy (including standard confocal microscopy) cannot provide spatial resolution greater than roughly one half the wavelength of the light observed. In practice, this means that two point sources of light emitting fluorescence simultaneously cannot be resolved if they lie closer than this resolution limit, known as the diffraction limit. Three general types of strategies have been developed to overcome this limitation. The first, known as stimulated emission depletion (STED) microscopy, scans samples with a beam of specially structured laser illumination. The second type uses 'stochastic' (random) switching of fluorescence of individual molecules to localize these molecules with a precision greater than the diffraction limit. The third uses sophisticated patterns of exciting light, known as structured illumination, to stimulate fluorescence from spatial regions smaller than the diffraction limit [140]. Below we will consider two representative methods of super-resolution microscopy, STED and a 'stochastic' method known as photoactivation light microscopy (PALM).

Stimulated emission/depletion microscopy (STED) uses a 'composite' spot of illumination that excites fluorescence from molecules lying within a small central region while it simultaneously suppresses the fluorescence of molecules lying more peripherally (Figure 24). As illustrated in the figure, the peripheral, fluorescence-suppressing component of the spot comprises light of a longer wavelength than the inner, fluorescence-exciting component. By increasing the intensity of the outer (longer-wavelength) component to high levels, the dimensions of the inner area, in which

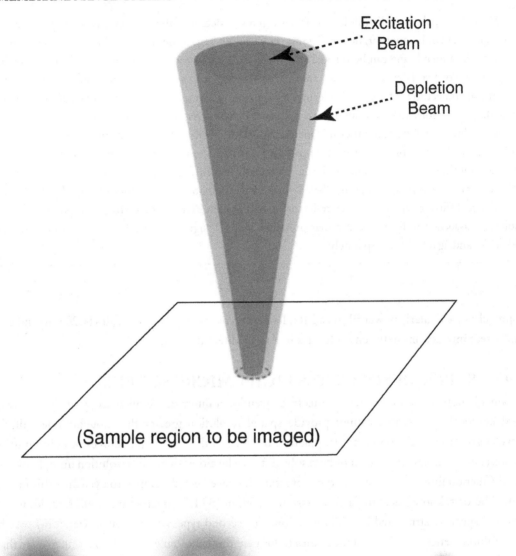

FIGURE 24: Stimulated emission-depletion (STED) microscopy uses nonlinear optical effects to achieve image resolution well below the diffraction limit. Through sophisticated optics, two laser beams are combined to produce a small region of illumination within the sample. As shown in the lower portion of the figure, illumination at the excitation wavelength is concentrated in the center of this region and surrounded by a 'depletion zone' that is illuminated at a different wavelength, suppressing normal sample fluorescence from the depletion zone. When the latter illumination is sufficiently strong, significant fluorescence will be emitted only from a central area much smaller than the diffraction limit (as small as 30 nm in diameter in biological applications). The focus of illumination is scanned across the entire region of interest to yield the final image.

---

fluorescence is excited, can be made much smaller than the diffraction limit. STED images are obtained by scanning the illuminating spot (in very small steps) across the sample region of interest, recording at each step the fluorescence emitted from the small volume illuminated by the inner region of the spot at that moment. To date, STED imaging has been applied to biological samples to achieve resolutions as high as 30–40 nm. A STED-based study [141] has provided evidence that in the plasma membrane, the lateral diffusion of sphingolipids and GPI-anchors, but not that of glycerophospholipids, exhibits brief (<20 ms) periods of confinement within tiny subregions of the membrane that are smaller than 20 nm in diameter.

Photoactivation localization microscopy (PALM) is an example of an alternative, 'stochastic' strategy for super-resolution fluorescence imaging. This method exploits the fact that an individual fluorescing molecule, when no nearby molecules are fluorescing, can be localized to a high degree of precision (much better than the diffraction limit) if a sufficiently large amount of emitted photons can be detected from the fluorescing species (Figure 25). Molecules to be localized by PALM are labeled with a photoactivatable fluorescent protein, which can be converted by light of one wavelength, λ_{act}, from a nonfluorescent species (PA) to an alternate form (PA*) that fluoresces when illuminated with light of a second wavelength, λ_{exc}. Imaging by PALM is based on many repetitions of a cycle of three steps, as illustrated in Figure 24. When a dense collection of PA-labeled molecules is briefly exposed to low-intensity light of wavelength λ_{act}, only a small fraction of the labeled molecules will be converted to their fluorescent (PA*-labeled) form. For PALM the intensity and duration of the activating light pulse are chosen so that all PA*-labeled molecules produced are separated from their nearest PA*-labeled neighbors by distances much greater than the diffraction limit. After activation as just described, the distribution of the PA*-labeled molecules is imaged (under illumination at λ_{exc}), for a period long enough for each photoactivated molecule to emit a large number of photons that allow it to be localized with high precision. At the end of this fluorescence-recording period, the PA*-labeled molecules are bleached by exposure to very strong

Steps in One PALM Cycle

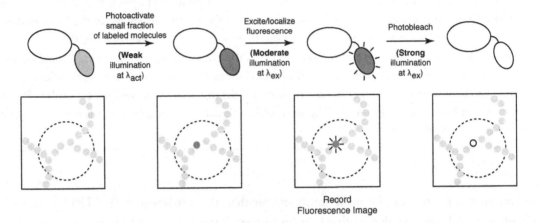

Compile Imaging Results from Many Cycles

FIGURE 25: Photoactivation light microscopy (PALM) localizes a few fluorescent molecules at a time, with high precision, to produce high-resolution images. As shown in the top and middle portions of the figure, in each PALM cycle, a few molecules of the protein of interest, 'tagged' with a photoactivatable protein, such as PA-GFP, are activated to a fluorescent form. The photoactivated molecules are then localized by fluorescence imaging, to a degree of spatial precision much better than the diffraction limit, and finally photobleached before the next activation/imaging cycle begins. Through many hundreds or thousands of such cycles, a composite image is created that maps the distribution(s) of the labeled species to high resolution (20–30 nm).

light at wavelength λ_{exc}, clearing the image field of fluorescent molecules. This basic cycle of activation, imaging and bleaching steps is repeated many times (hundreds or thousands), until most of the PA-labeled molecules in the image field have been activated and localized—all with a precision much better than the diffraction limit. Using current PALM methodology, images with resolutions as small as ca. 20 nm have been obtained from biological samples [142].

5.10   PLASMA MEMBRANE-DERIVED GIANT VESICLES

Treatment of cells with agents such as paraformaldehyde plus dithiothreitol, or prolonged incubations of cells in saline solutions, leads to 'blebbing' of large vesicles from the cell surface; the resulting vesicles may remain attached to the parent cells, but their membranes are largely detached from the cytoskeleton that is normally associated with the plasma membrane [99, 100, 143]. Interestingly, at temperatures below physiological (typically <25°C), blebbed vesicles exhibit segregation of large domains, which can readily be imaged by fluorescence microscopy when the membranes are labeled with fluorescent dyes that partition differentially between lipid phases with different degrees of order (Figure 26). This result demonstrates that the complex lipid composition of the plasma membrane, and the presence of large amounts of membrane proteins, does not inherently preclude the formation of large segregated domains qualitatively resembling liquid-ordered and liquid-disordered phases. However, experiments using fluorescence probes sensitive to lipid bilayer order suggest that the segregated domains formed in blebbed cell membrane vesicles differ less sharply in their degree of lipid ordering than do the segregated Lo and Ld domains observed in lipid model membranes [36]. This behavior may reflect a number of important differences between model and cellular membranes, including the more complex lipid composition and the high protein content of the latter.

While studies of blebbed plasma membrane vesicles have yielded interesting findings that may help to illuminate the organization of the plasma membrane in intact, unperturbed cells, it is clear that a number of properties of blebbed vesicles are significantly different from those of the 'native' plasma membrane. First, large segregated domains are typically not observed in unperturbed cell plasma membranes even at sub-physiological temperatures. Second, blebbed plasma membrane vesicles prepared by different treatments can differ quantitatively and sometimes even qualitatively

FIGURE 26: Large segregated domains with differing degrees of lipid order form in 'blebbed' plasma membrane vesicles at sub-physiological temperatures. Giant vesicles 'blebbed' from the plasma membrane of RBL cells, labeled with a fluorescent dye that preferentially labels liquid-disordered domains, were imaged at the indicated temperatures. Segregated domains of micron dimensions appear and enlarge as the temperature is progressively reduced below 25°C (reproduced with permission from reference [99]).

in properties such as the temperature dependence of visible domain segregation, the size and morphology of the segregated domains and the distributions of various membrane proteins between the coexisting domains [36, 99]. Some of these differences may be linked to chemical modifications of membrane proteins (e.g., depalmitoylation or crosslinking to membrane aminophospholipids) that occur using certain methods to induce blebbed vesicle formation [37, 144]. Third, the treatments used to prepare blebbed plasma membrane vesicles at least partially randomize the normal asymmetric distribution of different lipids between the two leaflets of the plasma membrane. Finally, mechanically generated projections from the plasma membrane, detached from the underlying cytoskeleton, have been shown to differ significantly from the unperturbed plasma membrane in their membrane protein composition [145], and the same may be true for blebbed plasma membrane vesicles. For these reasons, care should be applied in extrapolating the detailed features of the behavior of blebbed vesicles to the unperturbed plasma membrane.

5.11   LIPID MASS SPECTROMETRY (MS)

The past decade has seen dramatic improvements in both the sensitivity and the scope of mass-spectrometric analysis of membrane lipids, so that it is now possible to determine the levels of dozens or even hundreds of individual lipid molecular species in relatively small samples. This capability is critical to understand in detail the physical and the biological properties of membrane lipids, which can depend strongly on the structures of both a lipid's polar headgroup and its hydrocarbon chains. MS has also proven very useful to profile quantitatively the diversity of lipid species (notably including sphingolipids) in a variety of cell types, a capability of great importance given the increasing evidence that different molecular species of a given type of lipid (e.g., ceramide or sphingomyelin) can differ markedly in their signaling properties or their interactions with certain membrane proteins [23, 146].

The increasing sensitivity of membrane lipid analysis by MS offers a particular benefit for analysis of membrane preparations that are available only in very limited amounts, as is often the case for immunoisolated or other affinity-purified preparations of membranes from a particular cellular compartment or subcompartment. MS has, for example, been used to show [147] that in yeast, secretory vesicles derived from the trans-Golgi network (TGN) and bound for the plasma membrane are significantly enriched in sphingolipids and ergosterol (the major yeast sterol) compared to a TGN-enriched membrane fraction. This result suggests that sphingolipids and sterol become selectively enriched in these secretory vesicles during their formation at the TGN.

CHAPTER 6

Experimental Studies of Nanodomains in Cellular Systems

In the following sections, we will consider investigations of membrane nanodomains in several types of cellular systems that have been particularly useful, and widely studied, in building our current (limited) understanding of these often complex and sometimes elusive structures. Reflecting the historical development of the field, early investigations of many of these systems were framed in the context of the membrane raft model. Subsequent research into some of these systems has prompted considerable evolution in our models of the nature of raft domains. Investigations of other systems have revealed a variety of other physical mechanisms that can contribute to nanodomain formation, allowing generation of a wide range of nanodomains in membranes.

6.1 MEMBRANE DOMAIN ORGANIZATION IN FIBROBLASTS

Fibroblasts, because of their flattened morphology and ease of culture, have provided useful systems for microscopic studies of the lateral distribution and diffusion of diverse membrane components, including putative raft-associated molecules, within the plasma membrane. As noted earlier in this book, different raft-associated proteins appear homogeneously distributed in the plasma membrane at the resolution of fluorescence microscopy (200–250 nm), indicating that segregated raft domains in these cells, if present, must be smaller than this limit. Also as discussed earlier (Section 5), early fluorescence-microscopic studies found that when pairs of membrane components were independently crosslinked into patches using antibodies, various pairs of raft markers (GPI-proteins, influenza virus hemagglutinin or ganglioside GM1 bound to cholera toxin B-subunit) showed significant colocalization, while combinations of a 'raft' and a 'nonraft' marker did not [61, 104]. These findings were consistent with early models of rafts as membrane nanodomains that, though smaller than the optical resolution limit, are nonetheless large enough to incorporate multiple protein species, which can relocalize together when the distribution of these nanodomains is altered.

The early models of raft structure just noted prompted multiple studies that used FRET to investigate whether putative raft components are inhomogeneously distributed within cell plasma membranes (e.g., clustered or partitioned into 'raft' domains) on distance scales of the order of 10–20 nm. Results from early FRET studies of the distributions of different GPI-proteins in the plasma membrane were somewhat divergent but were generally consistent with a model in which a majority of these species is dispersed randomly within the membrane while a lesser fraction is present in very small clusters [129, 133]. Super-resolution imaging (PALM) has also provided evidence for formation of very small clusters of a GPI-anchored form of photoactivable GFP on the surface of cultured fibroblasts [127].

A very recent study using single-molecule fluorescence approaches [135] has shown that under 'basal' conditions, GPI-proteins such as CD59 exist as a mixture of monomers and homodimers, with the latter predominating at high expression levels, and small homo- and hetero-oligomers. Partial depletion of membrane cholesterol by cyclodextrin treatment weakened, but did not abolish, formation of GPI-protein homodimers and impaired formation of larger homo- and hetero-oligomers as well. These findings support recent models like those noted above, which suggest that under basal conditions raft protein components like GPI-proteins exist in the membrane as dispersed individual molecules or very small clusters/oligomers, possibly associating in a dynamic manner with small clusters of lipid molecules that exist in a relatively ordered state. Under appropriate conditions these small structures could in principle coalesce or expand to produce larger domains enriched in 'raft-philic' lipids and proteins, a possibility that is frequently invoked in recent models of rafts and their functions in signaling or trafficking processes ([92, 148]—see also the book by Kusumi et al. in this series). The findings just described also support current proposals that raft formation depends on a combination of lipid and protein-based interactions.

A variety of studies have sought to define the nature and origins of 'raft' domains in the cytoplasmic leaflet of the plasma membrane (and, potentially, of some other cellular membranes, e.g., in endosomes and the *trans*-Golgi). Formation of liquid-ordered domains in the cytoplasmic membrane leaflet is likely to rest on rather different physical bases than those that underlie formation of such domains in the extracellular leaflet, since sphingolipids are concentrated in the extracellular leaflet while the cytoplasmic leaflet contains largely unsaturated phospholipids. It is possible that some lipids present in the cytoplasmic leaflet of the membrane can functionally replace the contribution that sphingolipids make to formation of raft domains in the extracytoplasmic leaflet. More precise definition of lipid compositions of the two leaflets of the plasma membrane (and other cellular membranes), assisted by mass spectrometry among other methods, may help to assess this possibility. However, as noted in Section 2.1, studies with model systems also suggest that sphingolipid- and sterol-dependent segregation of lipids in the outer monolayer of the plasma membrane may be able to drive coupled segregation of lipids in the inner monolayer [9, 10]. Transmembrane proteins could in principle also promote coupled formation of raft nanodomains in the two leaflets of a given membrane. Various studies have reported that clustering or other redistributions of raft components in the outer leaflet of the plasma membrane can induce parallel redistribution of putative raft-associating species in the inner leaflet, suggesting that rafts can have a transmembrane character [105, 149]. Other studies, by contrast, have suggested that the distributions of raft nanodomains can be only weakly correlated between membrane leaflets [126, 150]. Much further work is clearly needed to understand the compositions and physical properties of rafts in the cytoplasmic leaflet of the plasma membrane and the degree to which rafts are coupled between the two leaflets of the membrane in various biological contexts.

6.2 NANOCLUSTER-BASED SIGNALING BY GPI- AND RAS PROTEINS

Current models for the potential functional importance of 'nanoclusters' of raft components arose in part from single-molecule studies that examined the dynamics and signaling properties of small clusters of GPI-proteins generated in the plasma membrane of unpolarized epithelial cells and fibroblasts. When multiple molecules of the GPI-protein CD59 bind on the cell surface to small gold particles, modified with low numbers of bound anti-CD59 antibody molecules, very small clusters of CD59 molecules form that recruit both the Src-family kinase Lyn and the heterotrimeric $G\alpha_{i2}$-protein (Figure 27 and [105, 107]). Association of both of the latter proteins with the clusters was transient, with average lifetimes for association of ca. 200 ms and 130 ms for Lyn and $G\alpha_{i2}$, respectively. Both the myristoylated/palmitoylated N-terminal 'anchor' sequence of Lyn (a 'raft-philic' structural motif) and other sequences within the protein promoted recruitment of Lyn to CD59 clusters, suggesting that both protein- and lipid-mediated interactions contribute to Lyn recruitment. $Gi_{\alpha2}$-mediated activation of Lyn in these clusters leads to transient immobilization of the clusters in the membrane, apparently through interactions with the submembrane actin cytoskeleton. Phosphatidylinositol-specific phospholipase $C\gamma2$ (PI-PLCγ2) was transiently recruited to these immobilized clusters, triggering Lyn-dependent activation of the phospholipase, consequent production of inositol trisphosphate (IP$_3$) from membrane-bound phosphatidylinositol-4,5-bisphosphate and local 'bursts' of IP$_3$-induced release of calcium from the ER. Through these molecular events, small-scale clustering of GPI-anchored CD59 thus prompts transient local recruitment and activation of signaling molecules that ultimately (by summation of many local calcium 'bursts') can contribute to a global intracellular response (calcium signaling), through processes in which protein- and lipid-mediated interactions as well as the actin cytoskeleton all play significant roles. The studies just discussed provide an instructive example of the complex and multifaceted molecular interactions that can underlie formation of membrane nanodomains and their signaling functions. Other studies [106, 151] have identified additional molecular components, including transmembrane proteins, that also contribute to the signaling and dynamic properties of nanoclusters of GPI-proteins generated at the cell surface.

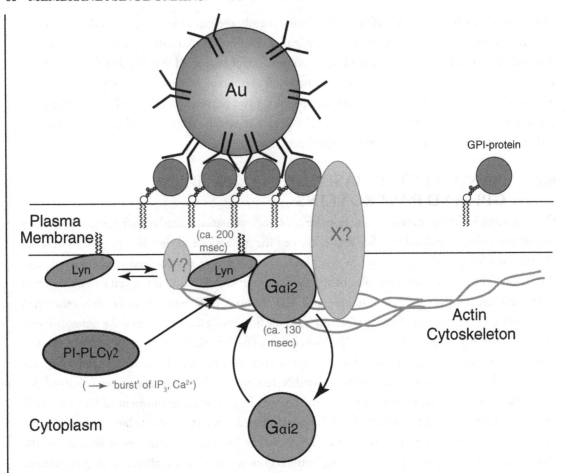

FIGURE 27: Clusters of GPI-proteins formed at the outer face of the plasma membrane recruit specific signaling proteins at the cytoplasmic face of the membrane. Antibody-gold nanoparticles that bind in a multivalent manner to GPI-proteins (e.g., CD59) at the plasma membrane generate GPI-protein microclusters (< 10 molecules), whose lateral diffusion is intermittently halted for short periods (< 1 s) through transient interactions with the submembrane actin cytoskeleton. The tyrosine kinase Lyn and the heterotrimeric G-protein subunit Gαi2 are transiently recruited to regions directly underlying CD59 microclusters, at the cytoplasmic face of the membrane. Recruitment of these proteins promotes transient arrest of microcluster diffusion and transient recruitment of the phospholipase PI-PLCγ2 to the plasma membrane. Membrane-recruited PI-PLCγ2 hydrolyzes plasma membrane PIP$_2$ (not shown), ultimately triggering a localized and transient elevation ('burst') of cytoplasmic calcium. Other proteins (like the hypothetical species shown as **X** and **Y** in the figure) could also participate in transiently coupling GPI-protein microclusters to the actin cytoskeleton.

Another well-studied group of plasma membrane signaling nanodomains in mammalian cells are those formed by the *ras* proteins. The classical *ras* proteins (H-, K- and N-*ras*) are small G-proteins that become activated upon stimulation of a number of cell-surface receptors, including the EGF receptor. In its resting (inactive) state, *ras* binds a molecule of GDP, which is replaced by GTP (in a reaction promoted by an appropriate *ras*-guanine nucleotide exchange factor [GEF], such as mSOS) to convert *ras* to its activated form. Subsequent deactivation of *ras* occurs through hydrolysis of the *ras*-bound GTP, promoted by a *ras*GTPase-activating protein(GAP), to return the *ras* molecule to its inactive state. In its activated state *ras* can recruit and activate a variety of down-stream effectors, including the Raf protein kinases and phosphatidylinositol-3-kinase.

While they are highly homologous in their guanine nucleotide-binding domains, the *ras* proteins differ greatly in their carboxyl-terminal hypervariable sequences, which comprise 23–24 contiguous amino acid residues ending in a lipid-modified membrane-binding sequence. H-*ras* is targeted to the plasma membrane through a carboxy-terminal farnesyl group and nearby palmi-toylated cysteine residues (lipid modifications of proteins are illustrated in Figure 5), while K-*ras* associates with this membrane via a carboxyl-terminal farnesyl residue and an adjacent polybasic amino acid sequence. Immunogold electron microscopy has shown that at the plasma membrane activated H- and K-*ras* form distinct nanoclusters and that 30–40% of each activated *ras* species is present in nanoclusters, independently of the level of *ras* activation [152, 153]. Increasing the number of activated H- or K-*ras* molecules in the plasma membrane produces a linear increase in the number of nanoclusters of the activated *ras* species. Galectin-1, which interacts preferentially with the GTP-bound form of H-*ras* [154], is present in nanoclusters of H-*ras* (GTP) and promotes their formation [155]. Galectin-3 appears to play a similar role in the formation of nanoclusters of activated K-*ras* [155]. Nanoclusters formed from activated H- or K-*ras* diffuse much more slowly in the membrane than do unclustered ras molecules, suggesting that these nanoclusters associate with additional membrane or cytoskeletal components that strongly restrict their mobility [156]. Consistent with this suggestion, disruption of cytoskeletal organization using the inhibitor latrunculin A reduces (though it does not abolish) formation of activated K-ras nanoclusters [157]. It is clear from such evidence that *ras* nanoclusters are generated through the interaction of a variety of membrane and juxtamembrane molecules.

The role of nanocluster formation in K-*ras* signaling has been studied extensively, using both *in vivo* and modeling-based approaches [158]. Nanoclusters, but not individual molecules, of activated K-*ras* recruit and activate the protein kinase Raf-1, which in turn activates the protein tyrosine kinase MEK; activated MEK then activates the MAP-kinases Erk1/2. Coordination of Raf-1 activation with successive activation of MEK and ERK1/2 is promoted by association of the three proteins with scaffolding proteins such as KSR (kinase suppressor of *ras*), adding another dimension to the postulated modular or 'switch-like' behavior of K-*ras* nanoclusters as signaling elements.

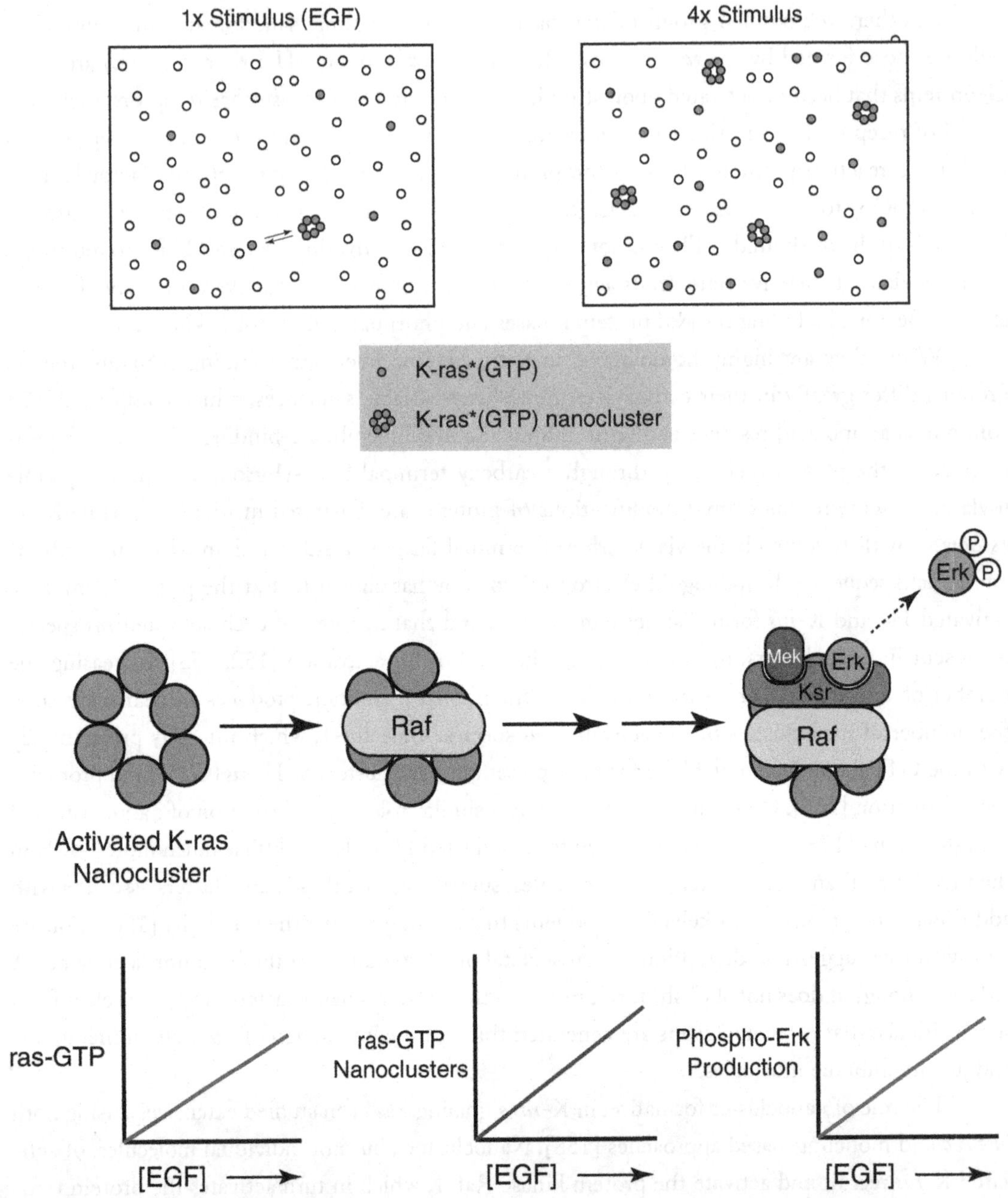

1x Stimulus (EGF)   4x Stimulus

○ K-ras*(GTP)

⚛ K-ras*(GTP) nanocluster

Activated K-ras
Nanocluster

Raf

Mek Erk

Ksr

Raf

Erk (P)(P)

ras-GTP

[EGF]

ras-GTP
Nanoclusters

[EGF]

Phospho-Erk
Production

[EGF]

FIGURE 28: Signaling through nanoclusters of activated ras enhances the signal/noise ratio of K-ras signaling while preserving its linear characteristics. Activated K-ras, generated by cell stimulation with epidermal growth factor (EGF), reversibly forms nanoclusters in equilibrium with activated K-ras monomers. The nanocluster density is linearly proportional to the overall level of activated K-ras present, which in turn varies linearly with the EGF concentration. Nanoclusters, but not monomers, of activated K-ras recruit Raf kinase and subsequently (via scaffolding proteins like KSR) the Mek and Erk1/2 kinases, ultimately producing phosphorylated Erk1/2. The average amount of phospho-Erk1/2 produced per nanocluster of activated K-ras is constant, and the overall output from K-ras signaling (i.e., total production of phospho-Erk1/2) thus varies linearly with the level of K-ras activation (and hence with the EGF concentration), over a wide range of EGF concentrations.

The average lifetime of a K-*ras* nanocluster (a few hundred milliseconds), and the average amount of activated Erk1/2 that each nanocluster produces during its lifetime, are thought to be constant regardless of the level of K-*ras* activation. Graded activation of K-*ras*, which as noted above leads to a directly proportional variation in the membrane density of activated K-*ras* nanoclusters, therefore should also lead to a linearly proportional variation in the total quantity of phospho-Erk1/2 produced within the cell (Figure 28). This prediction has been confirmed in experiments in which K-*ras* was activated to widely varying extents via graded stimulation of the EGF receptor [158]. This form of signal transmission, using *ras* nanoclusters as 'quantal' elements that nonetheless collectively produce a graded response to *ras* activation, may reduce 'noise' in K-*ras* signaling while preserving high sensitivity and linearity of signal transmission over a wide dynamic range [158].

6.3   ENDOCYTIC TRAFFICKING OF 'RAFT' COMPONENTS IN FIBROBLASTS

Clathrin-coated pits and caveolae provide two well-characterized examples of 'domains' that contribute to endocytic traffic from the plasma membrane [48]. Clathrin- and caveolae-dependent endocytosis are discussed in detail in the books by Lamaze and Stan in this series and will not be further discussed here. As well, however, a number of clathrin- and caveolin-independent endocytic pathways have also been discovered whose diversity can appear rather daunting [48, 159, 160]. Membrane rafts (which we here distinguish from caveolae) have been suggested to play a role in endocytic uptake of various components via some of these pathways, either at the level of cargo selection or at that of formation of the trafficking intermediates themselves. However, using current experimental methods it remains very challenging to demonstrate conclusively that rafts play a role in the endocytic trafficking of a given membrane component. We will briefly consider some of these challenges below.

One common experimental strategy to provide evidence for a role of rafts in a particular trafficking pathway, or in the trafficking of a particular membrane component, is to examine whether the process is affected by perturbing the levels of membrane cholesterol (most often, using cyclodextrins). A negative finding in such experiments may strongly suggest that rafts are not involved in the process of interest. However, positive findings must be interpreted with more caution, as discussed in Section 5.3. Membrane-trafficking processes typically rest on a complex and delicate orchestration of many different molecules and processes, and different trafficking pathways can exhibit regulatory 'cross-talk' (downregulating the activity of one endocytic pathway, for example, frequently leads to compensatory upregulation of others). Such trafficking processes may be particularly susceptible to 'off-target' perturbations when cellular cholesterol levels are perturbed with the aim of disrupting raft-dependent phenomena.

A potentially appealing approach to assess a role for rafts in trafficking of particular membrane components is to compare the trafficking of different species that are very closely related structurally and in most physical properties but differ in their affinity for rafts. The major challenge in such experiments is often to identify (or create) appropriate sets of molecules for comparison that fulfill the criterion just stated. This can be surprisingly difficult to accomplish in practice. A study by Ewers et al. [161] showed, for example, that ganglioside GM1 molecules with either long (C18) or short (C8) saturated N-acyl chains can both be incorporated into the plasma membranes of mouse melanoma cells and both serve as efficient receptors for cellular binding of the virus SV40, but that only the C18-GM1 species supported subsequent internalization of the virus. One plausible, and perhaps tempting, explanation for the failure of the C8-ganglioside to support viral internalization would be that the C18-GM1 is 'raft-philic' while the shorter-chain C8-GM1 is not. However, careful experimentation showed that the correct explanation is quite different: only the long-chain GM1 species is anchored to the membrane with sufficient mechanical strength to permit the bound virus to bend the cell membrane around itself, an essential step in viral uptake [161], and the difference just noted between long- and shorter-chain GM1 species is unrelated to their raft association.

As the preceding discussion illustrates, it can be quite challenging to demonstrate conclusively that rafts play a role in the trafficking of a given membrane component. We have recently used new experimental tools to examine three aspects of the endocytic trafficking of GPI-proteins in which rafts had been suggested to play a significant role based on previous experimental evidence [49, 162]. In only one of these three contexts, trafficking of GPI-proteins from the recycling endosome to the plasma membrane in CHO cells, did we obtain evidence for a functionally important role for rafts. There is clearly a pressing need for better and more stringent experimental tools to determine in which endocytic transport processes, and for which cargoes, rafts truly play a significant role. This task could become easier in some contexts if specific proteins, such as flotillins, prove to mediate truly raft-specific trafficking pathways. At present, however, no endocytic pathway has been shown to be raft-specific.

6.4   POLARIZED TRAFFICKING OF APICALLY LOCALIZED PROTEINS IN RENAL EPITHELIAL CELLS

As discussed earlier, in cells of the intestinal, renal and other polarized epithelia, the plasma membrane is differentiated into apical and basolateral domains, which exhibit distinct protein and lipid compositions and to which membrane components are delivered by distinct trafficking pathways. The mechanisms by which proteins are sorted in the Golgi complex for export to the basolateral membrane are comparatively well understood and often rest on protein–protein interactions [57]. By contrast, the mechanisms by which proteins are sorted in the Golgi for export to the apical surface appear to be more diverse and in many cases are less well understood. Multiple 'raft-philic' proteins, including diverse GPI-anchored proteins and the influenza hemagglutinin protein, are apically targeted in MDCK kidney epithelial cells, in which this phenomenon has been most intensively studied, and in various other epithelial cell lines (though not all). Diverse lines of evidence suggest that the affinity of these proteins for rafts may promote their apical trafficking from the trans-Golgi, though recent work has shown that additional targeting determinants also play a role in this process.

As discussed in Section 4, it was proposed a number of years ago that in MDCK and related epithelial cells, segregated lipid domains, enriched in sphingolipids and sterol, are found in the trans-Golgi network (TGN) and, along with 'raft-philic' protein cargo, become preferentially incorporated into transport carriers targeted to the apical plasma membrane [82, 83]. 'Proof of principle' for one important aspect of this proposal has recently been provided by findings that in budding yeast, immunoisolated post-Golgi transport vesicles destined for the plasma membrane are enriched in sphingolipids and ergosterol (relative to the parent Golgi compartment), indicating that lipid sorting can occur in the TGN [147]. Also consistent with the above proposal, glycosphingolipid synthesis has recently been shown to be essential for proper development of epithelial polarity in the intestine of *C. elegans* [119].

GPI-proteins were first proposed over two decades ago to undergo raft-dependent sorting in epithelial cells [163, 164], yet the mechanism by which these proteins are sorted for apical delivery in the Golgi/TGN remains only partially understood. GPI-anchored derivatives of 'reporter' proteins such as green fluorescent protein are apically targeted in MDCK cells, a finding that initially suggested that the GPI-anchor *per se* is sufficient as an apical targeting signal. However, subsequent evidence suggests that oligomerization is also a key requirement for apical sorting of GPI-proteins [165]. As well, it has been shown that fusing different GPI-anchor addition sequences to a given protein can alter both the polarity of targeting and the state of oligomerization (but not the 'raft-philic' character) of the resulting GPI-anchored species, presumably by directing protein attachment to structurally different GPI-anchors [166]. Taken together, these findings suggest that specific structural (and not solely physical) features of the GPI-anchor, a generic affinity of the GPI-anchor for ordered-lipid nanodomains and oligomerization of the protein may all play roles in the polarized targeting of GPI-proteins in epithelial cells.

The influenza virus hemagglutinin (HA) protein is also apically trafficked in epithelial cells. The protein was found to associate with isolated detergent-resistant membrane (DRM) fractions from these cells, suggesting that association of HA with rafts might be important to its apical targeting [167]. An essential apical targeting determinant was localized to the single transmembrane helix of the protein, specifically to the region spanning the extracytoplasmic membrane leaflet. A number of mutations in this region of the HA transmembrane helix inhibited both apical transport of HA and the normal association of the protein with isolated DRMs [167, 168], consistent with the proposal that association of HA with rafts is a key requirement for its apical sorting. However, some HA mutants could be isolated in DRMs but nonetheless showed impaired apical targeting. These results may imply, as the authors suggested, that raft association is a necessary but not a sufficient condition for apical targeting of the HA protein (noting again, however, the caveat that association of a given species with DRMs is not always a reliable indicator of its association with rafts in cell membranes (see Section 5.1)). The hypothesized additional determinants of apical targeting of the HA protein remain to be defined precisely.

Multiple membrane-associated proteins have been shown to be important for proper apical targeting of protein cargoes in MDCK and other epithelial cell lines [57]. Knockdown of expression of the MAL/Vps17 protein impairs trafficking of a variety of raft-associating proteins from the Golgi complex to the apical plasma membrane, without affecting protein trafficking to the basolateral plasma membrane, in MDCK cells [169, 170]. The MAL protein, a multispanning membrane protein that readily oligomerizes, associates with detergent-resistant membrane fractions isolated from MDCK cells, and fluorescence microscopy has showed that antibody-induced clusters of the protein in the plasma membrane accumulate raft markers such as GPI-proteins [171]. These findings, and observations that MAL is found in TGN-derived transport vesicles and in the apical recycling endosome [172], have led to proposals that MAL functions as a scaffolding protein promoting the coalescence of rafts and/or the incorporation of rafts into transport vesicles. The homologous MAL2 protein may mediate transcytotic delivery of raft proteins to the apical plasma membrane in polarized hepatocytes [173].

Multiple galectins appear to play important roles in the apical transport of different glycoprotein and glycolipid cargoes in polarized epithelial cells. Galectin-3 is required for selective apical trafficking of various glycosylated proteins, by a carbohydrate-dependent but raft-independent pathway, in MDCK cells and cultured enterocytes [174]. In cells with reduced galectin-3 expression, proteins normally transported to the apical surface by the pathway just noted are delivered to the basolateral surface instead. Knockdown of expression of galectin-4, which binds to both glycoproteins and 2-hydroxylacylated sulfatides (sulfogalactocerebrosides), impairs transport of raft-associating glycoproteins to the apical plasma membrane of enterocytes [175]. It has been suggested that interactions of galectin-4 both with sulfatides in rafts and with specific protein-linked oligosaccharide residues may contribute to the function of this protein in cargo sorting for apical transport. A similar mechanism of action has been proposed for galectin-9, which binds to specific oligosaccharides on both glycoproteins and the glycosphingolipid Forssman antigen and is essential for maintenance of apical polarity and for apical targeting of the influenza hemagglutinin in MDCK cells [176]. An important question concerning the roles of galectins in polarized transport in epithelial cells is the degree to which different galectins access different secretory compartments, since these proteins are not secreted via the classical ER/Golgi pathway. Evidence has been reported that galectin-1 may reach the Golgi [177]. It has not yet been determined whether other galectins potentially involved in protein secretion also can access compartments within the secretory pathway. Interestingly, reduction of galectin-4 expression causes mistargeted cargo proteins to accumulate in an unidentified post-Golgi intracellular compartment [178], suggesting that some galectins may play roles in sorting or packaging transport cargo in secretory-pathway compartments distal to the Golgi itself. Consistent with this possibility, galectin-3 has been shown to be present in subapical recycling endosomes, in which it may interact with newly synthesized glycoproteins that pass through this compartment en route to the apical plasma membrane [179].

6.5   ACTIVATION OF THE T-CELL RECEPTOR

Immune cells, and particularly T-lymphocytes, have been widely investigated as a possible example of an important role for rafts in cellular signaling. T-lymphocytes are activated physiologically through binding of T-cell receptors (TCRs) to appropriate antigenic peptides complexed to MHC (major histocompatibility complex) proteins on the surfaces of antigen-presenting cells. Activation of the TCR, which can be modulated by co-stimulation of other receptors on the T-cell surface, leads to rapid activation of tyrosine kinases including ZAP-70 at the cytoplasmic face of the plasma membrane, which phosphorylate adaptor proteins including LAT (linker of activation of T-cells) and SLP-76 (Figure 29). These adaptor proteins, once phosphorylated, serve as centers for assembly of protein complexes that activate further downstream signaling events, including elevation of cytoplasmic calcium levels, activation of transcription factors such as NFAT and ultimately, T-cell activation.

Interest in a potential role for 'rafts' in T-cell signaling was initially stimulated by observations that nonreceptor tyrosine kinases such as Lck and Fyn, which mediate important early phosphorylation events in TCR-initiated signaling, as well as the LAT adaptor protein could be isolated in a

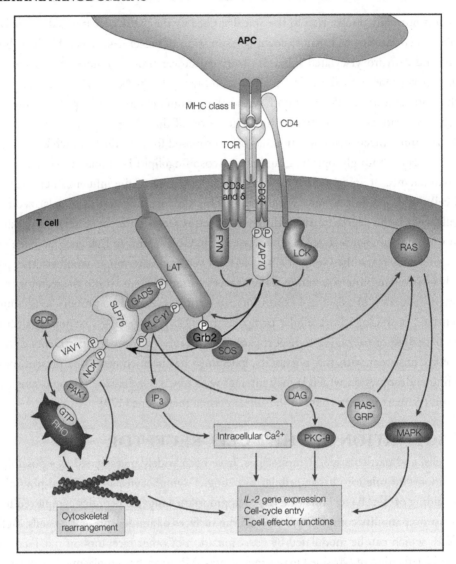

FIGURE 29: Schematic outline of signaling pathways linked to MHCII/antigen activation of the T-cell receptor. As shown, activation of the T-cell receptor stimulates activation of ZAP-70 mediated by the Lck and Fyn kinases. ZAP-70 in turn phosphorylates substrates including the scaffolding proteins LAT and SLP-76, which recruit a variety of additional downstream signaling molecules (adapted with permission from Abraham and Weiss, *Nat. Rev. Immunol.* 4 (2004), 301–308).

low-density, cold detergent-insoluble (DRM) fraction [180, 181]. It was also found that upon T-cell activation the TCR itself could be partly isolated in the DRM fraction, while membrane-associated signaling phosphatases such as CD45 were entirely excluded from this fraction in both resting and activated T-cells [182]. These observations led to early proposals that during T-cell activation signaling proteins such as Lck, LAT and activated TCR molecules associate with lipid rafts, promoting interactions among these proteins, 'sheltering' them from negative regulation by raft-excluded phosphatases and, through these two effects, promoting TCR-initiated T-cell activation. While (as noted already) association with DRMs is now considered a rather unreliable indicator of association of molecules with rafts in the living cell, other early lines of experimental evidence were also consistent with the hypothesis just outlined. In mutant T-cells lacking endogenous LAT or Lck, for example, normal TCR-initiated signaling could be restored by expressing wild-type LAT and Lck but not by expressing mutant forms of these proteins in which the membrane-anchoring motifs were altered from 'raft-philic' to 'raft-phobic' sequences [181, 182]. Interest in a possible role for rafts in T-cell activation was further stimulated by the finding that T-cells interact with antigen-presenting cells to form an extensive, structured region of contact known as an immunological synapse (discussed later in this section), and that raft components appeared to become enriched in the central region of this structure [183]. Subsequent studies indicated that early events in TCR-initiated signaling occur mainly in small TCR 'microclusters' rather than in the mature immunological synapse [184], shifting interest to understanding the organization and formation of these microclusters, and how rafts might contribute to their assembly and function.

While the possibility that rafts play major roles in T-cell signaling attracted great interest from immunologists, support for this proposal was never unanimous, in part because some experimental findings did not appear to support simple models for how rafts might contribute to T-cell activation (reviewed in [93]). For example, while some studies reported that fluorescent-labeled raft markers accumulate in the central region of the immunological synapse, others suggested that such observations simply reflect the highly folded topography of the plasma membrane in this region (for a discussion of how membrane topography can affect fluorescence imaging, see Section 5.2) and reported that raft markers are in fact not selectively enriched at the immunological synapse [109, 185]. Another influential study found that LAT and Lck associate extensively with other proteins in clusters during TCR-initiated signaling at the T-cell plasma membrane and that these associations are driven primarily by protein-mediated rather than by lipid-mediated interactions [186]. Other findings have also suggested that protein–protein interactions play a strong role in dictating the organization of LAT in the plasma membrane [35]. However, these findings leave open the possibility that rafts may play other roles in TCR-driven signaling, e.g., that exclusion of some proteins from rafts could be important for functionally segregating signal-propagating from signal-terminating proteins during the signaling process.

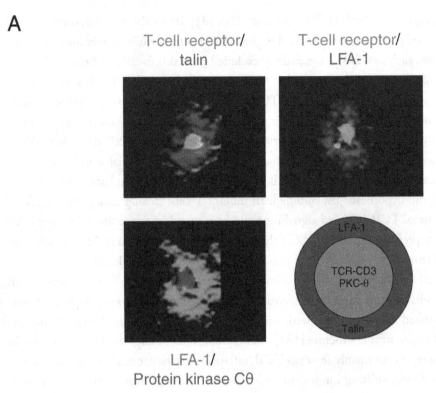

FIGURE 30: The 'immune synapse' formed between T-lymphocytes and antigen-presenting cells is spatially differentiated into regions whose protein composition is determined partly by steric factors. (A) The 'mature' immune synapse formed after an extended period of contact between a T-cell and an antigen-presenting cell (APC) is differentiated into a central cSMAC (central supramolecular activation cluster) and a surrounding pSMAC (peripheral supramolecular activation cluster). The T-cell receptor and various signaling molecules, including protein kinase C-θ as well as (not shown) Lck and ZAP-70, are enriched in the cSMAC. By contrast, the adhesion protein (integrin) LFA-1 and associated cytoskel-etal 'linker' proteins, including talin, are concentrated in the pSMAC. (B) Association of T-cell receptor and MHC molecules on the T-cell and APC surfaces, respectively, creates a relatively narrow separation of the cell surfaces in the cSMAC region. Proteins like CD45, whose extracellular domains extend far from the cell surface, are sterically excluded from this region. The intercellular adhesion complex formed by LFA-1 and ICAM-1 is also sterically excluded from the cSMAC region and segregates (along with associated species like talin) into the pSMAC, where the intercellular separation is greater. (Images in (A) reproduced with permission from Monks et al., *Nature* 395 [1998], 82–86; (B) reproduced with permission from reference [189]).

The mature immunological synapse formed between CD4 T-cells and antigen-presenting cells (APCs), the most extensively studied immunological synapse, exhibits two additional features that are particularly noteworthy in regard to potential mechanisms of membrane domain formation and the roles of nanodomains in T-cell signaling. The first such feature is the spatial organization of this immunological synapse and how it appears to be generated. As illustrated in Figure 30A, the synapse between a CD4 T-cell and an APC is spatially differentiated into a central region termed the cSMAC (central supramolecular activation cluster) and a surrounding annular region called the pSMAC (peripheral supramolecular activation cluster). In the T-cell plasma membrane, the cSMAC region is enriched in TCR molecules and associated signaling proteins, including the costimulatory receptors CD2 and CD28 and the Src-family protein kinases Lck and Fyn, while other proteins such as the adhesion protein LFA-1 are enriched in the pSMAC region [187]. Based on the narrow separation observed (roughly 13 nm) between the T-cell and APC surfaces in the cSMAC region, it was proposed that proteins that extend long distances from the cell surface would be sterically excluded from the cSMAC while proteins that extend shorter distances could be accom-modated in this region. This proposal is consistent with the known dimensions of the extracellular domains of several membrane molecules that are differentially distributed between the cSMAC and other regions of the T-cell membrane (see Figure 30B). The extracellular regions of the TCR, for example, and of the complex that it forms with MHC class I or class II on the APC surface could readily be accommodated within the narrow intercellular space found in the cSMAC, unlike that of

A

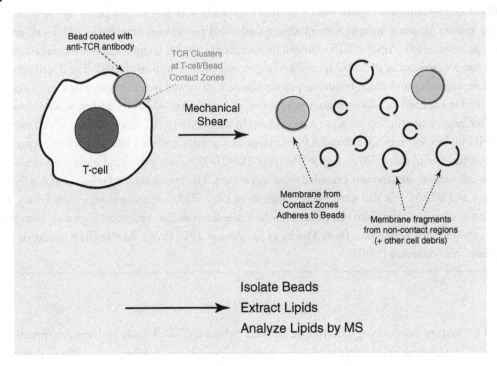

B

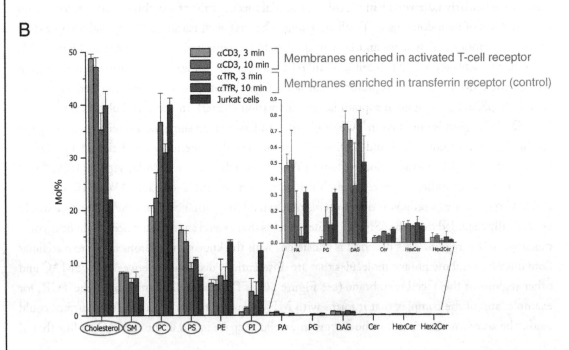

FIGURE 31: Regions of the T-cell plasma membrane enriched in activated T-cell receptors exhibit a distinctive lipid composition. (A) Schematic illustration of the immunoisolation of T-cell plasma membrane fragments enriched in (activated) T-cell receptor (TCR) molecules, using beads coupled to anti-TCR antibodies (198). Control preparations of plasma membrane fragments were also carried out (not shown) using beads coupled to antibodies against the transferrin receptor, a 'non-raft' plasma membrane protein. (B) Lipid analysis (using mass spectrometry) of the membrane fragments immunoisolated as illustrated in (A) using either anti-TCR or anti-(transferrin receptor) antibodies, from T-cells incubated with antibody-coupled beads for 3 min or 10 min. As shown by the highlighted portions of the bar graph, plasma membrane fragments enriched in activated TCR molecules are enriched in cholesterol, sphingomyelin (SM) and phosphatidylserine (PS), and depleted in phosphatidylcholine (PC) and phosphatidylinositol (PI), compared to plasma membrane fragments isolated using beads coupled to anti-(transferrin receptor) antibodies (reproduced with permission from reference [198]).

LFA-1, whose extracellular domain (alone or complexed to its binding partner, ICAM-1) extends a much greater distance from the surface.

Sterically based exclusion of some proteins from particular membrane domains has recently been suggested to occur, and to contribute importantly to T-cell signaling, in contexts other than the cSMAC alone. CD45, the most abundant protein tyrosine phosphatase in the T-cell plasma membrane, and some other tyrosine phosphatases that regulate TCR signaling possess extracellular domains that, like that of LFA-1, extend far from the cell surface. It was postulated [188], and subsequently demonstrated experimentally [184], that such phosphatases are excluded from TCR microclusters (also formed in regions of close T-cell/APC contact) in which TCR activation and signaling is initiated. A recent study [189] has shown that exclusion of CD45 from the vicinity of TCR molecules is essential to trigger MHC-stimulated TCR signaling and that steric factors are important for this exclusion. Two types of steric constraints appear to contribute to exclusion of CD45 from TCR microclusters: poor accommodation of the long, extended extracellular domain of CD45 within the narrow gap between the T-cell and APC surfaces, and lateral 'crowding' effects that promote exclusion of CD45 from microclusters densely packed with TCR and associated proteins [189, 190]. Interactions with the 'galectin lattice' and with the submembrane actin cytoskeleton may further modulate CD45 regulation of T-cell signaling [191], illustrating how multiple types of membrane domains and molecular interactions (attractive and repulsive) can contribute to the regulation of a membrane-associated signaling process.

A further noteworthy feature of the immunological synapse is that it is a dynamic structure shaped by multiple processes of active transport of membranes and membrane components. TCR microclusters, which form when T-cells bind to APCs bearing appropriate MHC-peptide complexes, rapidly become engaged by the cytoskeleton and are translocated toward the region of the cSMAC by retrograde actin flow [192]. Other components of the T-cell immunological synapse, such as LFA-1 bound to ICAM-1 on the APC surface, are also translocated centripetally in the same manner, in assemblies distinct from TCR microclusters [193]. Evidence suggests that the strength of coupling of different membrane protein clusters to the actin cytoskeleton during this process may influence the steady-state distribution of these proteins in the immunological synapse [192, 193]. The cSMAC is also a major center for polarized endo- and exocytic trafficking of the TCR and of other signaling molecules such as LAT, processes that strongly influence the intensity of TCR-mediated signaling within the cSMAC [194–197].

Two recent studies illustrate the potential of still-emerging technologies to extend our understanding of the roles of membrane nanodomains in T-cell activation. The first [198] used mass spectrometry to assess whether membrane rafts become concentrated in regions of TCR activation, using the strategy shown in Figure 31. Beads coated with a TCR-activating antibody were allowed to adsorb to human T-cells, clustering and activating TCR molecules in the regions of the cell surface in direct contact with the beads. After mechanically fragmenting the cells, the beads were isolated along with small attached fragments of the plasma membrane, derived from sites of cell/bead contact, whose lipid composition was then determined by mass spectrometry. For comparison, similar experiments were also carried out using beads coated with antibody against the transferrin receptor (TfR), a non-raft membrane protein that is not implicated in TCR signaling. Membrane fragments isolated using beads coated with anti-TCR antibody (and representing sites of TCR activation in the intact cells) were found to be enriched (compared to fragments isolated using anti-TfR-coated beads) in sphingomyelin and cholesterol, both species expected to be found in rafts. These findings suggest that raft components become enriched to a degree in regions of TCR activation.

Another recent paper [142] has used PALM to map the organization of several signaling molecules in the T-cell plasma membrane to a resolution of roughly 20 nm. In unstimulated cells LAT was found to exist mainly in very small clusters, and the extent of clustering increased significantly upon activation. By tagging different signaling proteins with photoactivatable fluorescent proteins, the authors showed that proteins such as PI-specific phospholipase C-γ and Grb2 became directly associated with LAT nanoclusters upon cell stimulation (Figure 32). Another downstream adaptor protein, SLP-76, showed a distinct distribution, with the protein associated with the periphery rather than the core of LAT clusters in activated cells. This distribution of SLP-76 was dependent on the integrity of actin microfilaments, underscoring once again the important role that the submembrane cytoskeleton can play in organizing membrane nanodomains. Another recent study using super-resolution microscopy [199] has shown that in T-cells activated Lck tends to self-associate into dynamic clusters, in which activated T-cell receptor molecules are also present while the protein tyrosine phosphatase CD45 is largely excluded. These findings illustrate the potential of new imaging methods, supported by quantitative analyses of molecular distributions, to examine

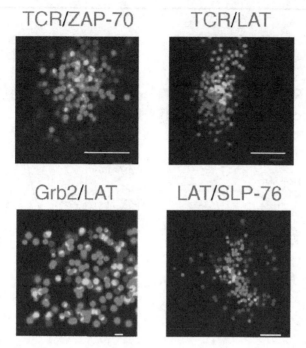

FIGURE 32: Super-resolution microscopy reveals distinct distributions of different components of T-cell receptor-initiated signaling at the plasma membrane. Each PALM image shows the relative distribution of two different proteins, both involved in early stages of T-cell receptor (TCR)-initiated signaling, at the plasma membrane in T-cells. All of these signaling proteins form clusters in the membrane of activated T-cells. The TCR and the ZAP-70 protein kinase co-cluster, consistent with the known direct binding of ZAP-70 to the TCR upon cell activation. LAT and the adaptor protein Grb2 also co-cluster, consistent with the known direct interaction of these proteins, in activated cells. By contrast, clusters of LAT and TCR are found in proximity to each other but show only limited overlap. SLP-76 appears to concentrate immediately outside clusters of LAT molecules (reproduced with permission from reference [142]).

the nanoscale organization of molecules at the plasma membrane in unprecedented detail [200]. As our understanding of the spatial and temporal organization of signaling nanodomains in T-cells becomes more precise, theoretical modeling will undoubtedly become increasingly useful to understand the functional implications of such organization, as it has in the case of *ras* signaling.

6.6   MAST CELLS AND THE FcεRI RECEPTOR

The role of membrane organization in immune signaling has also been investigated extensively for the high-affinity IgE receptor (FcεRI) in mast cells. This receptor is activated physiologically

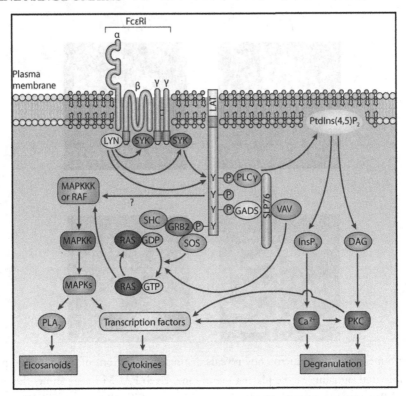

FIGURE 33: Schematic illustration of molecules and pathways involved in signaling initiated by activation of the mast cell receptor (FcεRI). Binding of antigen-IgE complexes to the mast cell receptor triggers enhanced tyrosine phosphorylation of the receptor by the kinase Lyn, promoting binding of the Syk kinase to the receptor. Lyn and Syk then phosphorylate LAT, which recruits a variety of additional proteins mediating downstream signaling events (reproduced with permission from Gilfillan and Tkaczyk, *Nat. Rev. Immunol.* 6 (2006), 218–230).

by binding of multivalent antigens to IgE molecules bound to FcεRI molecules on the mast cell surface. Studies with artificial antigens of varying valency, geometry and size have revealed that receptor activation depends on antigen-induced recruitment of FcεRI into clusters within which neighboring FcεRI molecules can approach to separations of a few nanometers. Receptor clustering leads to rapid phosphorylation of immunoreceptor tyrosine-based activation motifs (ITAMs) in the cytoplasmic tails of FcεRI subunits, mediated by the nonreceptor kinase Lyn. Tyrosine-phosphorylated FcεRI molecules recruit additional molecules of Lyn as well as the tyrosine kinase Syk, which phosphorylates downstream targets (including the two isoforms of LAT) to initiate further signaling events (Figure 33 and [201]).

Most studies of the role of membrane nanodomain structure in mast cell signaling have focused on the earliest phase of signaling, the phosphorylation of FcεRI's ITAM motifs following antigen-induced clustering of the receptor. Despite much investigation, the detailed mechanism of this process is still not well understood. Low levels of Lyn kinase appear to associate with FcεRI molecules even in unactivated mast cells, through an interaction independent of ITAM phosphorylation. Antigen-induced clustering of FcεRI molecules, a small fraction of which are associated with Lyn, is thought to promote Lyn-mediated phosphorylation of multiple receptor molecules within the same cluster, promoting further Lyn recruitment to the cluster and stimulating other early signaling events. Negative regulation of tyrosine phosphatases may also play a role in promoting early stages of receptor-initiated cell activation [202] and undoubtedly contributes to setting the 'poise' of the cell signaling apparatus in the resting state to avoid cell activation under inappropriate conditions.

Interest in a potential role for lipid nanodomains in FcεRI-mediated signaling was first prompted by findings that a portion of the total cellular pool of both FcεRI and Lyn kinase were associated with detergent-resistant low-density membrane (DRM) fractions isolated from activated mast cells, while DRM fractions isolated from unactivated mast cells contained Lyn but no FcεRI [203]. The kinetics of appearance of phosphorylated FcεRI in DRM fractions paralleled those of receptor activation. It was proposed that antigen-induced receptor clustering promoted association of the receptor with lipid rafts in which Lyn was already present (and receptor-deactivating phosphatases potentially depleted), thereby promoting efficient and sustained phosphorylation of the receptor ITAM motifs and consequent downstream signaling events. Since detergent solubility can be an unreliable indicator of raft association, additional experimental approaches have been explored to assess the mesoscale distribution of FcεRI and associated signaling molecules during mast cell activation. One group of studies used immunogold labeling and electron microscopy to compare the distributions of FcεRI and other signaling molecules in the plasma membrane of mast cells [204, 205]. In resting cells, a fraction of plasma membrane FcεRI was found in small patches, with which a small but significant fraction of Lyn was associated. In activated cells Lyn was still observed intermixed with FcεRI in small- or medium-size clusters of the latter but was associated only with the periphery of large FcεRI clusters; in contrast, Syk was recruited to the core regions of the large FcεRI clusters. Upon cell activation the adaptor protein LAT shifted from a randomly dispersed to a more clustered distribution; LAT clusters did not overlap with clusters of FcεRI, but a significant fraction of LAT clusters was found in close proximity to clusters of FcεRI. The nanoscale organization of these signaling molecules is in some ways reminiscent of that found in the PALM studies of T-cell signaling discussed earlier, with some proteins freely intermixed within clusters of a given protein species while other proteins are found only in the peripheral portions of such clusters, or in distinct but contiguous clusters.

The results just discussed do not directly address the question of possible raft association of FcεRI and related signaling molecules in mast cells, but they indicate that association between different molecules involved in FcεRI signaling cannot be driven simply by random partitioning into common raft domains. This conclusion is reinforced by findings that in the mast cell plasma membrane some pairs of membrane raft markers, such as the GPI-protein Thy-1 and ganglioside GM1, cluster in distinct domains when they are independently crosslinked [206]. The distributions (and redistributions) of various membrane proteins during FcεRI-mediated mast cell activation are therefore likely to be determined to an important degree by protein–protein interactions, though in principle lipid-mediated interactions may play a significant role as well. Evidence has also been reported suggesting an important role for the actin cytoskeleton in determining the association of proteins such as Lyn with FcεRI clusters at the inner surface of the plasma membrane [207]. As in other examples discussed already, FcεRI–initiated signaling nanodomains are clearly complex structures whose assembly rests on multiple types of intermolecular interactions.

In parallel with the functional and microscopic studies discussed above, mast cells have also been extensively studied for evidence of the presence of liquid-ordered domains in the plasma membrane, and for possible reorganization when FcεRI is activated by antigens. Electron spin resonance spectroscopy of membrane lipid probes, which can detect even nanoscale inhomogeneities in lipid organization, has indicated that in isolated mast cell plasma membrane vesicles the majority of the membrane lipids exhibits properties resembling those of a liquid-ordered phase, while a smaller fraction of the lipids exhibits properties similar to those of a liquid-disordered phase [208]. It has also been shown that fluorescent resonance energy transfer (see Section 5.6) between donor and acceptor lipid probes in the plasma membrane of intact mast cells is more efficient when both probes carry saturated hydrocarbon chains, and hence are both 'raft-philic,' than when one species is 'raft-philic' while the other is 'raft-phobic' [209]. These FRET observations suggest that the mast cell plasma membrane shows a heterogeneous lipid lateral organization, with coexisting domains possibly as small as a few nm in size comprising more vs. less ordered lipids.

As the above results illustrate, experimental findings to date have led to an increasingly complex picture of the organization of various effector and regulatory proteins with respect to FcεRI clusters in the plasma membrane of resting and antigen-activated mast cells. The mesoscale organization of membrane lipids may play a significant role in FcεRI-initiated activation of mast cells, but the precise nature of this role remains to be defined, as does the detailed organization of lipid nanodomains in the plasma membrane of resting and activated cells.

• • • •

CHAPTER 7

Looking Forward

It is now clear that a wide variety of molecular interactions, with diverse physical origins, spatial ranges and degrees of specificity, work together in varying combinations to generate mesoscale 'inhomogeneities' (clustering, confinement, exclusion, etc.) in the distributions of many cell membrane components. It is also clear that such phenomena play highly important roles in many aspects of membrane function. The already-established principles and examples discussed in this book amply illustrate the potentially vast range of types of nanodomains that could be present in a given biological membrane. However, for this very reason the concept of 'nanodomains' may prove to be fundamentally inadequate to describe the mesoscale organization of many membrane proteins and lipids. Many membrane components may participate simultaneously in multiple, spatially overlapping networks or 'webs' of interactions within the membrane, and for many purposes it may be more illuminating to consider membranes as organized into 'nanopatterns,' formed by distinct but overlapping sets of membrane components, than into discrete 'nanodomains.'

As this book has sought to make clear, there remains a great deal that we do *not* know concerning the mesoscale organization of cell membranes; we therefore will end by noting a few of the major current challenges that this field must address. First, we still need to fully catalogue the 'elements,' i.e., the general types of molecular interactions and mechanisms, that contribute to formation of domains within membranes. It is already clear that these elements can be quite diverse, ranging from highly specific protein–protein interactions and trafficking pathways to relatively nonspecific (and often underappreciated) factors such as steric repulsions, long-range electrostatic interactions and the responses of membranes to short-and long-range mechanical stresses. It is equally clear that our inventory of these elements is not yet complete, as new possibilities are reported regularly (see for example [78] and [193]). Second, we currently know much more about the mesoscale organization of cell plasma membranes (though even for these membranes our knowledge is quite rudimentary) than that of most other cellular membranes. Third, while the interactions of integral membrane proteins with lipids have been investigated for decades using biophysical and structural–biological methods, to date very few such studies have examined membrane proteins thought to be associated with rafts or related membrane domains. Similarly, at present we still lack high-resolution structures for the membrane proteins implicated in formation of many known (and postulated) types of membrane domains. Finally, we need to understand more fully the temporal characteristics of membrane nanodomains. As we have seen, functionally important nanoclusters of membrane molecules like the *ras* proteins exhibit lifetimes of only hundreds of milliseconds, and other molecular interactions implicated in nanodomain formation may exhibit lifetimes much shorter still (135; see also the book by Kusumi et al. in this series). Single-molecule imaging offers one promising avenue to investigate such short-lived phenomena, but understanding some aspects of membrane nanodomain dynamics may require additional methods that combine high spatial resolution with even greater (sub-millisecond) temporal resolution.

Happily, all of the challenges just noted can be addressed, in many cases using existing (if not always widely available) technologies. We can anticipate that studies of the mesoscale organization of membranes, and of the functional importance of this organization, will remain an exciting 'frontier' area of cell biology for some time to come.

. . . .

References

[1]   Shevchenko, A. and Simons, K. (2010) Lipidomics: coming to grips with lipid diversity. *Nat. Rev. Mol. Cell Biol.* 11, pp. 593–8.

[2]   Thompson, T. E. and Tillack, T. W. (1985) Organization of glycosphingolipids in bilayers and plasma membranes of mammalian cells. *Annu. Rev. Biophys. Chem.* 14, pp. 361–86.

[3]   Gennis, R. B. (1989) Biomembranes: molecular structure and function. New York: Springer-Verlag.

[4]   Brown, D. A. and London, E. (1998) Structure and origin of ordered lipid domains in biological membranes. *J. Membr. Biol.* 164, pp. 103–14.

[5]   Feigenson, G. W. (2006) Phase behavior of lipid mixtures. *Nat. Chem. Biol.* 2, pp. 560–3.

[6]   Silvius, J. R. (2003) Fluorescence energy transfer reveals microdomain formation at physiological temperatures in lipid mixtures modeling the outer leaflet of the plasma membrane. *Biophys. J.* 85, pp. 1034–45.

[7]   Honerkamp-Smith, A. R., Veatch, S. L. and Keller, S. L. (2009) An introduction to critical points for biophysicists: observations of compositional heterogeneity in lipid membranes. *Biochim. Biophys. Acta* 1788, pp. 53–63.

[8]   Feigenson, G. W. (2009) Phase diagrams and lipid domains in multicomponent lipid bilayer mixtures. *Biochim. Biophys. Acta* 1788, pp. 47–52.

[9]   Wan, C., Kiessling, V. and Tamm, L. K. (2008) Coupling of cholesterol-rich lipid phases in asymmetric bilayers. *Biochemistry* 47, pp. 2190–8.

[10] Collins, M. D. and Keller, S. L. (2008) Tuning lipid mixtures to induce or suppress domain formation across leaflets of unsupported asymmetric bilayers. *Proc. Natl. Acad. Sci. USA* 105, pp. 124–8.

[11] Wang, T.-Y. and Silvius, J. R. (2001) Cholesterol does not induce segregation of liquid-ordered domains in bilayers modeling the inner leaflet of the plasma membrane. *Biophys. J.* 81, pp. 2762–73.

[12] Hammond, A. T., Heberle, F. A., Baumgart, T., Holowka, D., Baird, B. and Feigenson, G. W. (2005) Crosslinking a lipid raft component triggers liquid ordered–liquid disordered phase separation in model plasma membranes. *Proc. Natl. Acad. Sci. USA* 102, pp. 6320–5.

[13]   White, S. W. (2009) Biophysical dissection of membrane proteins. *Nature* 459, pp. 344–6.

[14]   Reichow, S. L. and Gonen, T. (2009) Lipid–protein interactions probed by electron crystallography. *Curr. Op. Struc. Biol.* 19, pp. 560–5.

[15]   Ulmschneider, M. B., Sansom, M. S. P. and Di Nola, A. (2005) Properties of integral membrane protein structures: derivation of an implicit membrane potential. *Prot. Struc. Function Bioinf.* 59, pp. 252–65.

[16]   Marsh, D. (2008) Protein modulation of lipids, and vice-versa, in membranes. *Biochim. Biophys. Acta* 1778, pp. 1545–75.

[17]   Marsh, D. and Horváth. L. I. (1998) Structure, dynamics and composition of the lipid–protein interface. Perspectives from spin-labelling. *Biochim. Biophys. Acta* 1376, pp. 267–96.

[18]   Lee A. G. (2003) Lipid–protein interactions in biological membranes: a structural perspective. *Biochim. Biophys. Acta* 1612, pp. 1–40.

[19]   Takamori, S., Holt, M., Stenius, K., Lemke, E. A., Grønborg, M., Riedel, D., Urlaub, H., Schenck, S., Brügger, B., Ringler, P., Müller, S. A., Rammner, B., Gräter, F., Hub, J. S., De Groot, B. L., Mieskes, G., Moriyama, Y., Klingauf, J., Grubmüller, H., Heuser, J., Wieland, F. and Jahn, R. (2006) Molecular anatomy of a trafficking organelle. *Cell* 127, pp. 831–46.

[20]   Dupuy, A. D. and Engelman, D. M. (2008) Protein area occupancy at the center of the red blood cell membrane. *Proc. Natl. Acad. Sci. USA* 105, pp. 2848–52.

[21]   Hite, R. K., Li, Z. and Walz, T. (2010) Principles of membrane protein interactions with annular lipids deduced from aquaporin-0 2D crystals. *EMBO J.* 29, pp. 1652–8.

[22]   Qin, L., Sharpe, M. A., Garavito, R. M. and Ferguson-Miller, S. (2007) Conserved lipid-binding sites in membrane proteins: a focus on cytochrome c oxidase. *Curr. Op. Struc. Biol.* 17, pp. 444–50.

[23]   Contreras, F.-X., Ernst, A. M., Haberkant, P., Björkholm, P., Lindahl, E., Gönen, B., Tischer, C., Elofsson, A., von Heijne, G., Thiele, C., Pepperkok, R., Wieland, F. and Brügger, B. (2012) Molecular recognition of a single sphingolipid species by a protein's transmembrane domain. *Nature* 481, pp. 525–9.

[24]   Barrett P. J., Song, Y., Van Horn, W. D., Hustedt, E. J., Schafer, J. M., Hadziselimovic, A., Beel, A. J. and Sanders, C. R. (2012) The amyloid precursor protein has a flexible transmembrane domain and binds cholesterol. *Science* 336, pp. 1168–71.

[25]   Lemmon, M. A. (2008) Membrane recognition by phospholipid-binding domains. *Nat. Rev. Mol. Cell Biol.* 9, pp. 99–112.

[26]   Smith, D. C., Lord, J. M., Roberts, L. M. and Johannes, L. (2004) Glycosphingolipids as toxin receptors. *Sem. Cell Dev. Biol.* 15, pp. 397–408.

[27]   Wang, J., Gambhir, A., Hangyás-Mihályné, G., Murray, D., Golebiewska, U. and Mc-

Laughlin, S. (2002) Lateral sequestration of phosphatidylinositol 4,5-bisphosphate by the basic effector domain of myristoylated alanine-rich C kinase substrate is due to nonspecific electrostatic interactions. *J. Biol. Chem.* 277, pp. 34401–12.

[28] Golebiewska, U., Gambhir, A., Hangyás-Mihályné, G., Zaitseva, I., Rädler, J. and Mc-Laughlin, S. (2006) Membrane-bound basic peptides sequester multivalent (PIP$_2$), but not monovalent (PS), acidic lipids. *Biophys. J.* 91, pp. 588–99.

[29] Dietrich, C., Volovyk, Z. N., Levi, M., Thompson, N. L. and Jacobson, K. (2001) Partitioning of Thy-1, GM1, and cross-linked phospholipid analogs into lipid rafts reconstituted in supported model membrane monolayers *Proc. Natl. Acad. Sci. USA* 98, pp. 10642–7.

[30] Kahya, N., Brown, D. A. and Schwille, P. (2005) Raft partitioning and dynamic behavior of human placental alkaline phosphatase in giant unilamellar vesicles. *Biochemistry* 44, pp. 7479–89.

[31] Wang, T.-Y., Leventis, R. and Silvius, J. R. (2001) Partitioning of lipidated peptide sequences into liquid-ordered lipid domains in model and biological membranes. *Biochemistry* 40, pp. 13031–40.

[32] Brown, D. A. (2006) Lipid rafts, detergent-resistant membranes, and raft targeting signals. *Physiology* 21, pp. 430–9.

[33] Saad, J. S., Miller, J., Tai, J., Kim, A., Ghanam, R. H. and Summers, M. F. (2006) Structural basis for targeting HIV-1 Gag proteins to the plasma membrane for virus assembly. *Proc. Natl. Acad. Sci. USA* 103, pp. 11364–9.

[34] Fastenberg, M. E., Shogomori, H., Xu, X., Brown, D. A. and London, E. (2003) Exclusion of a transmembrane-type peptide from ordered-lipid domains (rafts) detected by fluorescence quenching: extension of quenching analysis to account for the effects of domain size and domain boundaries. *Biochemistry* 42, pp. 12376–90.

[35] Shogomori, H., Hammond, A. T., Ostermeyer-Fay, A. G., Barr, D. J., Feigenson, G. W., London, E. and Brown, D. A. (2005) Palmitoylation and intracellular domain interactions both contribute to raft targeting of linker for activation of T cells. *J. Biol. Chem.* 280, pp. 18931–42.

[36] Kaiser, H. J., Lingwood, D., Levental, I., Sampaio, J. L., Kalvodova, L., Rajendran, L. and Simons, K. (2009) Order of lipid phases in model and plasma membranes. *Proc. Natl. Acad. Sci. USA* 106, pp. 16645–50.

[37] Levental, I., Lingwood, D., Grzybek, M., Coskun, U. and Simons, K. (2010) Palmitoylation regulates raft affinity for the majority of integral raft proteins. *Proc. Natl. Acad. Sci. USA* 107, pp. 22050–4.

[38] Antonny, B. (2011) Mechanisms of membrane curvature sensing. *Annu. Rev. Biochem.* 80, pp. 101–23.

[39]   Suetsugu S., Toyookac, K. and Senjua, Y. (2010) Subcellular membrane curvature mediated by the BAR domain superfamily proteins. *Sem. Cell Dev. Biol.* 21, pp. 340–9.

[40]   Galic, M., Jeong, S., Tsai, F.-C., Joubert L.M., Wu, Y. I., Hahn, K. M., Cui, Y. and Meyer, T. (2012) External push and internal pull forces recruitcurvature-sensing N-BAR domain proteins to the plasma membrane. *Nat. Cell Biol.* 14, pp. 874–81.

[41]   Shibata, Y., Hu, J., Kozlov, M. M. and Rapoport, T. A. (2009) Mechanisms shaping the membranes of cellular organelles. *Annu. Rev. Cell Dev. Biol.* 25, pp. 329–54.

[42]   Stachowiak, J. C., Schmid, E. M., Ryan, C. J., Ann, H. S., Sasaki, D. Y., Sherman, M. B., Geissler, P. L., Fletcher, D. A. and Hayden, C. C. (2012) Membrane bending by protein-protein crowding. *Nat. Cell Biol.* 14, pp. 944–9.

[43]   Čopič, A., Latham, C. F., Horlbeck, M. A., D'Arcangelo, J. G. and Miller, E. A. (2012) ER cargo properties specify a requirement for COPII coat rigidity mediated by Sec13p. *Science* 335, pp. 1359–62.

[44]   Tian, A. and Baumgart, T. (2009) Sorting of lipids and proteins in membrane curvature gradients. *Biophys. J.* 96, pp. 2676–88.

[45]   Roux, A., Cuvelier, D., Nassoy, P., Prost, J., Bassereau, P. and Goud, B. (2005) Role of curvature and phase transition in lipid sorting and fission of membrane tubules. *EMBO J.* 24, pp. 1537–45.

[46]   Sorre, B., Callan-Jones, A., Manneville, J. B., Nassoy, P., Joanny, J. F., Prost, J., Goud, B. and Bassereau, P. (2009) Curvature-driven lipid sorting needs proximity to a demixing point and is aided by proteins. *Proc. Natl. Acad. Sci. USA* 106, pp. 5622–6.

[47]   Sabharanjak, S., Sharma, P., Parton, R. G. and Mayor, S. (2002) GPI-anchored proteins are delivered to recycling endosomes via a distinct cdc42-regulated, clathrin-independent pinocytic pathway. *Dev. Cell* 2, pp. 411–23.

[48]   Mayor, S. and Pagano, R. E. (2007) Pathways of clathrin-independent endocytosis. *Nat. Mol. Cell Biol.* 8, pp. 603–12.

[49]   Bhagatji, P., Leventis, R., Comeau, J., Refaei, M. and Silvius, J. R. (2009) Steric and not structure-specific factors dictate the endocytic mechanism of glycosylphosphatidylinositol-anchored proteins. *J. Cell. Biol.* 186, pp. 615–28.

[50]   Pearse, B. M. F. and Bretscher, M. F. (1981) Membrane recycling by coated vesicles. *Annu. Rev. Biochem.* 50, pp. 85–101.

[51]   Voeltz, G. K., Prinz, W. A., Shibata, Y., Rist, J. M. and Rapoport, T. A. (2006) A class of membrane proteins shaping the tubular endoplasmic reticulum. *Cell* 124, pp. 573–86.

[52]   Shibata, Y., Shemesh, T., Prinz, W. A., Palazzo, A. F., Kozlov, M. M. and Rapoport, T. A. (2010) Mechanisms determining the morphology of the peripheral ER. *Cell* 143, pp. 774–88.

[53] Charrin, S., Le Naour, F., Silvie, O., Milhiet, P.-E., Boucheux, C. and Rubinstein, E. (2009) Lateral organization of membrane proteins: tetraspanins spin their web. *Biochem. J.* 420, pp. 133–54.

[54] Min, G., Wang, H., Sun, T. T. and Kong, X. P. (2006) Structural basis for tetraspanin functions as revealed by the cryo-EM structure of uroplakin complexes at 6-Å resolution. *J. Cell. Biol.* 173, pp. 975–83.

[55] Yáñez-Mó, M., Barreiro, O., Gordon-Alonso, M., Sala-Valdé, M. and Sánchez-Madrid, F. (2009) Tetraspanin-enriched microdomains: a functional unit in cell plasma membranes. *Trends Cell Sci.* 19, pp. 434–46.

[56] Otto, G. P. and Nichols, B. J. (2011) The roles of flotillin microdomains-endocytosis and beyond. *J. Cell. Sci.* 124, pp. 3933–40.

[57] Weisz, O. A. and Rodriguez-Boulan, E. (2009) Apical trafficking in epithelial cells: signals, clusters and motors. *J. Cell. Sci.* 122, pp. 4253–66.

[58] Antón, O. M., Andrés-Delgado, L., Reglero-Real, N., Batista, A. and Alonso, M. A. (2011) MAL protein controls protein sorting at the supramolecular activation cluster of human T lymphocytes. *J. Immunol.* 186, pp. 6345–56.

[59] Ramnarayanan, S. P. and Tuma, P. L. (2011) MAL, but not MAL2, expression promotes the formation of cholesterol-dependent membrane domains that recruit apical proteins. *Biochem. J.* 439, pp. 497–504.

[60] Rothberg, K. G., Heuser, J. E., Donzell, W. C., Ying, Y. S., Glenney, J. R. and Anderson, R. G. (1992) Caveolin, a protein component of caveolae membrane coats. *Cell* 68, pp. 673–82.

[61] Mayor, S., Rothberg, K. G. and Maxfield, F. R. (1994) Sequestration of GPI-anchored proteins in caveolae triggered by cross-linking. *Science* 264, pp. 1948–51.

[62] Schnitzer, J. E., McIntosh, D. P., Dvorak, A. M., Liu, J. and Oh, P. (1995) Separation of caveolae from associated microdomains of GPI-anchored proteins. *Science* 269, pp. 1435–9.

[63] Oh, P. and Schnitzer, J. E. (2001) Segregation of heterotrimeric G proteins in cell surface microdomains. G_q binds caveolin to concentrate in caveolae, whereas G_i and G_s target lipid rafts by default. *Mol. Biol. Cell* 12, pp. 685–98.

[64] Zheng, Y. Z., Boscher, C., Inder, K. L., Fairbank, M., Loo, D., Hill, M. M., Nabi, I. R. and Foster, L. J. (2011) Differential impact of caveolae and caveolin-1 scaffolds on the membrane raft proteome. *Mol. Cell Proteomics* 10, p. M110.007146.

[65] Lajoie, P., Goetz, J. G., Dennis, J. W. and Nabi, I. R. (2009) Lattices, rafts, and scaffolds: domain regulation of receptor signaling at the plasma membrane. *J. Cell Biol.* 185, pp. 381–5.

[66] Dennis, J. W., Nabi, I. R. and Demetriou, M. (2009) Metabolism, cell surface organization, and disease. *Cell* 139, pp. 1229–41.

[67]   Lajoie, P., Partridge, E. A., Guay, G., Goetz, J. G., Pawling, J., Lagana, A., Joshi, B., Dennis, J. W. and Nabi, I. R. (2007) Plasma membrane domain organization regulates EGFR signaling in tumor cells. *J. Cell Biol.* 171, pp. 341–56.

[68]   Dennis, J. W., Lau, K. S., Demetriou, M. and Nabi, I. R. (2009) Adaptive regulation at the cell surface by N-glycosylation. *Traffic* 10, pp. 1569–78.

[69]   Golebiewska, U., Kay, J. G., Masters, T., Grinstein, S., Im, W., Pastor, R. W., Scarlata, S. and McLaughlin, S. (2011) Evidence for a fence that impedes the diffusion of phosphatidylinositol 4,5-bisphosphate out of the forming phagosomes of macrophages. *Mol. Biol. Cell* 22, pp. 3498–507.

[70]   Lee, W. L., Mason, D., Schreiber, A. D. and Grinstein, S. (2007) Quantitative analysis of membrane remodeling at the phagocytic cup. *Mol. Biol. Cell* 18, pp. 2883–92.

[71]   Estey, M. P., Kim, M. S. and Trimble, W. S. (2011) *Septins. Curr. Biol.* 21, pp. R384–7.

[72]   Hu, Q., Milenkovic, L., Jin, H., Scott, M. P., Nachury, M. V., Spiliotis, E. T. and Nelson, W. J. (2010) A septin diffusion barrier at the base of the primary cilium maintains ciliary membrane protein distribution. *Science* 329, pp. 436–9.

[73]   Chih, B., Liu, P., Chinn, Y., Chalouni, C., Komuves, L. G., Hass, P. E., Sandoval, W. and Peterson, A. S. (2012) A ciliopathy complex at the transition zone protects the cilia as a privileged membrane domain. *Nat. Cell Biol.* 14, pp. 61–72.

[74]   Funamoto, S., Meili, R., Lee, S., Parry, L. and Firtel, R. A. (2002) Spatial and temporal regulation of 3-phosphoinositides by PI 3-kinase and PTEN mediates chemotaxis. *Cell* 109, pp. 611–23.

[75]   Nishio, M., Watanabe, K.-I., Sasaki, J., Taya, C., Takasuga, S., Iizuka, R., Balla, T., Yamazaki, M., Watanabe, H., Itoh, R., Kuroda, S., Horie, Y., Förster, I., Mak, T. W., Yonekawa, H., Penninger, J. M., Kanaho, Y., Suzuki, A. and Sasaki, T. (2007) Control of cell polarity and motility by the PtdIns(3,4,5)P3phosphatase SHIP1. *Nat. Cell Biol.* 9, pp. 36–44.

[76]   Lavi, Y., Edidin, M. A. and Gheber, L. A. (2007) Dynamic patches of membrane proteins. *Biophys. J.* 93, pp. L35–7.

[77]   Lavi, Y., Gov, N., Edidin, M. and Gheber, L. A. (2012) Lifetime of major histocompatibility complex class-I membrane clusters is controlled by the actin cytoskeleton. *Biophys. J.* 102, pp. 1543–50.

[78]   Gowrishankar, K., Ghosh, S., Saha, S., Mayor C. R., S. and Rao, M. (2012) Active remodeling of cortical actin regulates spatiotemporal organization of cell surface molecules. *Cell* 149, pp. 1353–67.

[79]   Singer, S. J. and Nicolson, G. L. (1972) The fluid mosaic model of the structure of cell membranes. *Science* 175, pp. 720–31.

[80]   Ipsen, J. H., Karlstrom, G., Mouritsen, O. G., Wennerstrom, H. and Zuckermann, M. J.

(1987) Phase equilibria in the phosphatidylcholine-cholesterol system. *Biochim. Biophys. Acta.* 905, pp. 162–72.

[81] Vist, M. R. and Davis, J. H. (1990) Phase equilibria of cholesterol/dipalmitoylphosphati-dylcholine mixtures: ^{2}H nuclear magnetic resonance and differential scanning calorimetry studies. *Biochemistry* 29, pp. 451–64.

[82] van Meer, G. and Simons, K. (1988) Lipid sorting in epithelial cells. *Biochemistry* 27, pp. 6197–202.

[83] van Meer, G. and Simons, K. (1988) Lipid polarity and sorting in epithelial cells. *J. Cellular Biochem.* 36, pp. 51–58.

[84] Brown, D. A. and Rose, J. (1992) Sorting of GPI-anchored proteins to glycolipid-enriched membrane subdomains during transport to the apical cell surface. *Cell* 68, pp. 533–44.

[85] Schroeder, R., London, E. and Brown, D. A. (1994) Interactions between saturated acyl chains confer detergent resistance on lipids and glycosylphosphatidylinositol (GPI)-anchored proteins: GPI-anchored proteins in liposomes and cells show similar behavior. *Proc. Natl. Acad. Sci.* USA 91, pp. 12130–4.

[86] Schuck, S., Honsho, M., Ekroos, K., Shevchenko, A. and Simons, K. (2003) Resistance of cell membranes to different detergents. *Proc. Natl. Acad. Sci. USA* 100, pp. 5795–800.

[87] Brown, D. A. and London, E. (1997) Structure of detergent-resistant membrane domains: does phase separation occur in biological membranes? *Biochem. Biophys. Res. Commun.* 240, pp. 1–7.

[88] Simons, K. and Ikonen, E. (1997) Functional rafts in cell membranes. *Nature* 387, pp. 569–72.

[89] Brown, D. A. and London, E. (1998) Functions of lipid rafts in biological membranes. *Annu. Rev. Cell Dev. Biol.* 14, pp. 111–36.

[90] Brown, D. A. and London, E. (2000) Structure and function of sphingolipid- and choles-terol-rich membrane rafts. *J. Biol. Chem.* 275, pp. 17221–4.

[91] Munro, S. (2003) Lipid rafts: elusive or illusive? *Cell* 115, pp. 377–88.

[92] Hancock, J. F. (2006) Lipid rafts: contentious only from simplistic standpoints. *Nat. Rev. Mol. Cell Biol.* 7, pp. 456–62.

[93] Shaw, A. S. (2006) Lipid rafts: now you see them, now you don't. *Nat. Immunol.* 7, pp. 1139–42.

[94] Kenworthy, A. K. (2008) Have we become overly reliant on lipid rafts? Talking Point on the involvement of lipid rafts in T-cell activation. *EMBO Rept.* 9, pp. 531–5.

[95] He, H.-T. and Marguet, D. (2008) T-cell antigen receptor triggering and lipid rafts: a mat-ter of space and time scales. Talking Point on the involvement of lipid rafts in T-cell activa-tion. *EMBO Rept.* 9, pp. 536–40.

[96]   Macdonald, J. L. and Pike, L. J. (2005) A simplified method for the preparation of detergent-free lipid rafts. *J. Lipid Res.* 46, pp. 1061–7.

[97]   Dietrich, C., Bagatolli, L. A., Volovyk, Z. N., Thompson, N. L., Levi, M., Jacobson, K. and Gratton, E. (2000) Lipid rafts reconstituted in model membranes. *Biophys. J.* 80, pp. 1417–28.

[98]   Veatch, S. L., Soubias, O., Keller, S. L. and Gawrisch, K. (2007) Critical fluctuations in domain-forming lipid mixtures. *Proc. Natl. Acad. Sci. USA* 104, pp. 17650–5.

[99]   Baumgart, T., Hammond, A. T., Sengupta, P., Hess, S. T., Holowka, D. A., Baird, B. A. and Webb, W. W. (2008) Large-scale fluid/fluid phase separation of proteins and lipids in giant plasma membrane vesicles. *Proc. Natl. Acad. Sci. USA* 104, pp. 3165–70.

[100]   Veatch, S. L., Cicuta, P., Sengupta, P., Honerkamp-Smith, A., Holowka, D. and Baird, B. (2008) Critical fluctuations in plasma membrane vesicles. *ACS Chem. Biol.* 3, pp. 287–93.

[101]   Heerklotz, H. (2002) Triton promotes domain formation in lipid raft mixtures. *Biophys. J.* 83, pp. 2693–701.

[102]   Pathak, P. and London, E. (2011) Measurement of lipid nanodomain (raft) formation and size in sphingomyelin/POPC/cholesterol vesicles shows TX-100 and transmembrane helices increase domain size by coalescing preexisting nanodomains but do not induce domain formation. *Biophys. J.* 101, pp. 2417–25.

[103]   Song, K. S., Li, S., Okamoto, T., Quilliam, L. A., Sargiacomo, M. and Lisanti, M. P. (1996) Co-purification and direct interaction of Ras with caveolin, an integral membrane protein of caveolae microdomains. Detergent-free purification of caveolae membranes. *J. Biol. Chem.* 271, pp. 9690–7.

[104]   Harder, T., Scheiffele, P., Verkade, P. and Simons, K. (1998) Lipid domain structure of the plasma membrane revealed by patching of membrane components. *J. Cell Biol.* 141, pp. 929–42.

[105]   Suzuki, K. G., Fujiwara, T. K., Edidin, M. and Kusumi, A. (2007) Dynamic recruitment of phospholipase C gamma at transiently immobilized GPI-anchored receptor clusters induces IP_3-Ca^{2+} signaling: single-molecule tracking study 2. *J. Cell Biol.* 177, pp. 731–42.

[106]   Chen, Y., Thelin, W. R., Yang, B., Milgram, S. L. and Jacobson, K. (2006) Transient anchorage of cross-linked glycosyl-phosphatidylinositol-anchored proteins depends on cholesterol, Src family kinases, caveolin, and phosphoinositides. *J. Cell Biol.* 175, pp. 169–78.

[107]   Suzuki, K. G., Fujiwara, T. K., Sanematsu, F., Iino, R., Edidin, M. and Kusumi, A. (2007) GPI-anchored receptor clusters transiently recruit Lyn and G alpha for temporary cluster immobilization and Lyn activation: single-molecule tracking study 1. *J. Cell. Biol.* 177, pp. 717–30.

[108]   van Rheenen, J., Achame, E. M., Janssen, H., Calafat, J. and Jalink, K. (2005) PIP_2 signaling in lipid domains: a critical re-evaluation. *EMBO J.* 24, pp. 1664–73.

[109] Glebov, O. O. and Nichols, B. J. (2004) Lipid raft proteins have a random distribution during localized activation of the T-cell receptor. *Nat. Cell Biol.* 6, pp. 238–43.

[110] Gaus, K., Chklovskaia, E., Fazekas de St Groth, B., Jessup, W. and Harder, T. (2005) Condensation of the plasma membrane at the site of T lymphocyte activation. *J. Cell Biol.* 171, pp. 121–31.

[111] Hao, M., Lin, S. X., Karylowski, O. J., Wustner, D., McGraw, T. E. and Maxfield, F. R. (2002). Vesicular and non-vesicular sterol transport in living cells. The endocytic recycling compartment is a major sterol storage organelle. *J. Biol. Chem.* 277, pp. 609–17.

[112] Wustner, D. (2007) Plasma membrane sterol distribution resembles the surface topography of living cells. *Mol. Biol. Cell* 18, pp. 211–28.

[113] Subtil, A., Gaidarov, I., Kobylarz, K., Lampson, M. A., Keen, J. H. and McGraw, T. E. (1999) Acute cholesterol depletion inhibits clathrin-coated pit budding. *Proc. Natl. Acad. Sci. USA* 96, pp. 6775–80.

[114] Pizzo, P., Giurisato, E., Tassi, M., Benedetti, A., Pozzan, T. and Viola, A. (2002) Lipid rafts and T cell receptor signaling: a critical re-evaluation. *Eur. J. Immunol.* 32, pp. 3082–91.

[115] Shvartsman, D. E., Gutman, O., Tietz, A. and Henis, Y. I. (2006) Cyclodextrins but not compactin inhibit the lateral diffusion of membrane proteins independent of cholesterol. *Traffic* 7, pp. 917–26.

[116] Wang, J., Megha and London, E. (2004) Relationship between sterol/steroid structure and participation in ordered lipid domains (lipid rafts): implications for lipid raft structure and function. *Biochemistry* 43, pp. 1010–8.

[117] Lahiri, S. and Futerman, A. H. (2007) The metabolism and function of sphingolipids and glycosphingolipids. *Cell. Mol. Life Sci.* 64, pp. 2270–84.

[118] Furukawa, K., Tokuda, N., Okuda, T., Tajima, O. and Furukawa, K. (2004) Glycosphingolipids in engineered mice: insights into function. *Sem. Cell Dev. Biol.* 15, pp. 389–96.

[119] Zhang, H., Abraham, N., Khan, L. A., Hall, D. H., Fleming, J. T. and Gobel, V. (2011) Apicobasal domain identities of expanding tubular membranes depend on glycosphingolipid biosynthesis. *Nat. Cell Biol.* 13, pp. 1189–201.

[120] Proszynski, T. J., Klemm, R. W., Gravert, M., Hsu, P. P., Gloor, Y., Wagner, J., Kozak, K., Grabner, H., Walzer, K., Bagnat, M., Simons, K. and Walch-Solimena, C. (2005) A genome-wide visual screen reveals a role for sphingolipids and ergosterol in cell surface delivery in yeast. *Proc. Natl. Acad. Sci. USA* 102, pp. 17981–6.

[121] Gaigg, B., Toulmay, A. and Schneiter, R. (2006) Very long-chain fatty acid-containing lipids rather than sphingolipids per se are required for raft association and stable surface transport of newly synthesized plasma membrane ATPase in yeast. *J. Biol. Chem.* 281, pp. 34135–45.

[122] Bacia, K., Schuette, C. G., Kahya, N., Jahn, R. and Schwille, P. (2004) SNAREs prefer

liquid-disordered over "raft" (liquid-ordered) domains when reconstituted into giant unilamellar vesicles. *J. Biol. Chem.* 279, pp. 37951–5.

[123] Wang, T.-Y., Leventis, R. and Silvius, J. R. (2000) Fluorescence-based evaluation of the partitioning of lipids and lipidated peptides into liquid-ordered lipid microdomains: a model for molecular partitioning into "lipid rafts." *Biophys. J.* 79, pp. 919–33.

[124] Wang, T.-Y., Leventis, R. and Silvius, J. R. (2001) Partitioning of lipidated peptide sequences into liquid-ordered lipid domains in model and biological membranes. *Biochemistry* 40, pp. 13031–40.

[125] Rothberg, K. G., Ying, Y. and Koller, J. F. (1990) The glycophospholipid-linked folate receptor internalizes folate without entering the clathrin-coated pit endocytic pathway. *J. Cell. Biol.* 110, pp. 637–49.

[126] Prior, I. A., Muncke, C., Parton, R. G. and Hancock, J. F. (2003) Direct visualization of Ras proteins in spatially distinct cell surface microdomains. *J. Cell. Biol.* 160, pp. 165–70.

[127] Sengupta, P., Jovanovic-Talisman, T., Skoko, D., Renz, M., Veatch, S. L. and Lippincott-Schwartz, J. (2011) Probing protein heterogeneity in the plasma membrane using PALM and pair correlation analysis. *Nat. Methods* 8, pp. 969–75.

[128] Boyle, S., Kolin, D. L., Bieler, J. G., Schneck, J. P., Wiseman, P. W. and Edidin, M. (2011) Quantum dot fluorescence characterizes the nanoscale organization of T cell receptors for antigen. *Biophys. J.* 101, pp. L57–9.

[129] Kenworthy, A. K. and Edidin, M. (1998) Distribution of a glycosylphosphatidylinositol-anchored protein at the apical surface of MDCK cells examined at a resolution of 100 Å using imaging fluorescence resonance energy transfer. *J. Cell. Biol.* 142, pp. 69–84.

[130] Kenworthy, A. K., Petranova, N. and Edidin, M. (2000) High-resolution FRET microscopy of cholera toxin B-subunit and GPI-anchored proteins in cell plasmamembranes. *Mol. Biol. Cell* 11, pp. 1645–55.

[131] Zacharias, D. A., Violin, J. D., Newton, A. C. and Tsien, R. Y. (2002) Partitioning of lipid-modified monomeric GFPs into membrane microdomains of live cells. *Science* 296, pp. 913–6.

[132] Varma, R. and Mayor, S. (1998) GPI-anchored proteins are organized in submicron domains at the cell surface. *Nature* 394, pp. 798–801.

[133] Sharma, P., Varma, R., Sarasij, R. C., Ira, Gousset, K., Krishnamoorthy, G., Rao, M. and Mayor, S. (2004) Nanoscale organization of multiple GPI-anchored proteins in living cell membranes. *Cell* 116, 5 pp. 77–89.

[134] Kerppola, T. K. (2008) Bimolecular fluorescence complementation (BiFC) analysis as a probe of protein interactions in living cells. *Annu. Rev. Biophys.* 37, pp. 465–87.

[135] Suzuki, K. G. N., Kasai, R. S., Hirosawa, K. M., Nemoto, Y. L., Ishibashi, M., Miwa, Y.,

Fujiwara, T. K. and Kusumi, A. (2012) Transient GPI-anchored protein homodimers are units for raft organization and function. *Nat. Chem. Biol.* 8, pp. 774–83.

[136] He, J., Yu, T., Pan, J. and Li, H. (2012) Visualisation and identification of the interaction between STIM1s in resting cells. *PLoS One* 7, p. e33377.

[137] Saxton, M. J. and K. Jacobson. (1997) Single-particle tracking: applications to membrane dynamics. *Annu. Rev. Biophys. Biomol. Struct.* 26, pp. 373–99.

[138] Mattheyses, A. L., Simon, S. M. and Rappoport, J. Z. (2010) Imaging with total internal reflection fluorescence microscopy for the cell biologist. *J. Cell Sci.* 123, pp. 3621–8.

[139] Kasai, R. S., Suzuki, K. G., Prossnitz, E. R., Koyama-Honda, I., Nakada, C., Fujiwara, T. K. and Kusumi, A. (2011) Full characterization of GPCR monomer–dimer dynamic equilibrium by single molecule imaging. *J. Cell Biol.* 192, pp. 463–80.

[140] Shao, L., Kner, P., Rego, E. H. and Gustafsson, M. G. (2011) Super-resolution 3D microscopy of live whole cells using structured illumination. *Nat. Methods* 12, pp. 1044–6.

[141] Eggeling, C., Ringemann, C., Medda, R., Schwarzmann, G., Sandhoff, K., Polyakova, S., Belov, V. N., Hein, B., von Middendorff, C., Schonle, A. and Hell, S. W. (2009) Direct observation of the nanoscale dynamics of membrane lipids in a living cell. *Nature* 457, pp. 1159–62.

[142] Sherman, E., Barr, V., Manley, S., Patterson, G., Balagopalan, L., Akpan, I., Regan, C. K., Merrill, R. K., Sommers, C. L., Lippincott-Schwartz, J. and Samelson, L. E. (2011) Functional nanoscale organization of signaling molecules downstream of the T cell antigen receptor. *Immunity* 35, pp. 705–20.

[143] Lingwood, D., Ries, J., Schwille, P. and Simons, K. (2008) Plasma membranes are poised for activation of raft phase coalescence at physiological temperature. *Proc. Natl. Acad. Sci. USA* 105, pp. 10005–10.

[144] Levental, I., Grzybek, M. and Simons, K. (2011) Raft domains of variable properties and compositions in plasma membrane vesicles. *Proc. Natl. Acad. Sci. USA* 108, pp. 11411–6.

[145] Discher, D. E, Mohandas, N. and Evans, E. A. (1994) Molecular maps of red cell deformation: hidden elasticity and in situ connectivity. *Science* 266, pp. 1032–5.

[146] Hannun, Y. A. and Obeid, L. M. (2011) Many ceramides. *J. Biol. Chem.* 286, pp. 27855–62.

[147] Klemm, R. W., Ejsing, C. S., Surma, M. A., Kaiser, H. J., Gerl, M. J., Sampaio, J. L., de Robillard, Q., Ferguson, C., Proszynski, T. J., Shevchenko, A. and Simons, K. (2009) Segregation of sphingolipids and sterols during formation of secretory vesicles at the trans-Golgi network. *J. Cell. Biol.* 185, pp. 601–12.

[148] Simons, K. and Gerl, M. (2010) Revitalizing membrane rafts: new tools and insights. *Nat. Rev. Mol. Cell Biol.* 11, pp. 688–99.

[149]   Pyenta, P. S., Holowka, D. and Baird, B. (2001) Cross-correlation analysis of inner-leaflet-anchored green fluorescent protein co-redistributed with IgE receptors and outer leaflet lipid raft components. *Biophys. J.* 80, pp. 2120–32.

[150]   Wu, M., Holowka, D., Craighead, H. G. and Baird, B. (2004) Visualization of plasma membrane compartmentalization with patterned lipid bilayers. *Proc. Natl. Acad. Sci. USA* 101, pp. 13798–803.

[151]   Chen, Y., Veracini, L., Benistant, C. and Jacobson, K. (2009) The transmembrane protein CBP plays a role in transiently anchoring small clusters of Thy-1, a GPI-anchored protein, to the cytoskeleton. *J. Cell Sci.* 122, pp. 3966–72.

[152]   Abankwa, D., Gorfe, A. A. and Hancock, J. F. (2007) Ras nanoclusters: molecular structure and assembly. *Sem. Cell Dev. Biol.* 18, pp. 599–607.

[153]   Henis, Y. I., Hancock, J. F. and Prior, I. A. (2009) Ras acylation, compartmentalization and signaling nanoclusters. *Mol. Membr. Biol.* 26, pp. 80–92.

[154]   Belanis, L., Plowman, S. J., Rotblat, B., Hancock, J. F. and Kloog, Y. (2008) Galectin-1 is a novel structural component and a major regulator of h-ras nanoclusters. *Mol. Biol. Cell* 19, pp. 1404–14.

[155]   Tian, T., Plowman, S. J., Parton, R. G., Kloog, Y. and Hancock, J. F. (2010) Mathematical modeling of K-Ras nanocluster formation on the plasma membrane. *Biophys. J.* 99, pp. 534–43.

[156]   Murakoshi, H., Iino, R., Kobayashi, T., Fujiwara, T., Ohshima, C., Yoshimura, A. and Kusumi, A. (2004) Single-molecule imaging analysis of Ras activation in living cells. *Proc. Natl. Acad. Sci. USA* 101, pp. 7317–22.

[157]   Plowman, S. J., Muncke, C., Parton, R. G. and Hancock, J. F. (2005) H-ras, K-ras, and inner plasma membrane raft proteins operate in nanoclusters with differential dependence on the actin cytoskeleton. *Proc. Natl. Acad. Sci. USA* 102, pp. 15500–5.

[158]   Tian, T., Harding, A., Inder, K., Plowman, S., Parton, R. G. and Hancock, J. F. (2007) Plasma membrane nanoswitches generate high-fidelity Ras signal transduction. *Nat. Cell Biol.* 9, pp. 905–14.

[159]   Howes, M. T., Kirkham, M., Riches, J., Cortese, K., Walser, P. J., Simpson, F., Hill, M. M., Jones, A., Lundmark, R., Lindsay, M. R., Hernandez-Deviez, D. J., Hadzic, G., Mc-Cluskey, A., Bashir, R., Liu, L., Pilch, P., McMahon, H., Robinson, P. J., Hancock, J. F., Mayor, S. and Parton, R. G. (2010) Clathrin-independent carriers form a high capacity endocytic sorting system at the leading edge of migrating cells. *J. Cell. Biol.* 190, pp. 675–91.

[160]   Sandvig, K., Pust, S., Skotland, T. and van Deurs, B. (2011) Clathrin-independent endocytosis: mechanisms and function. *Curr. Op. Cell Biol.* 23, pp. 413–20.

[161]   Ewers, H., Romer, W., Smith, A. E., Bacia, K., Dmitrieff, S., Chai, W., Mancini, R.,

Kartenbeck, J., Chambon, V., Berland, L., Oppenheim, A., Schwarzmann, G., Feizi, T., Schwille, P., Sens, P., Helenius, A. and Johannes, L. (2009) GM1 structure determines SV40-induced membrane invagination and infection. *Nat. Cell Biol.* 12, pp. 11–8.

[162] Refaei, M., Leventis, R. and Silvius, J. R. (2011) Assessment of the roles of ordered lipid microdomains in post-endocytic trafficking of glycosyl-phosphatidylinositol-anchored proteins in mammalian fibroblasts. *Traffic* 12, pp. 1012–24.

[163] Brown, D. A., Crise, B. and Rose, J. K. (1989) Mechanism of membrane anchoring affects polarized expression of two proteins in MDCK cells. *Science* 245, pp. 1499–501.

[164] Lisanti, M. P., Caras, I. W., Davitz, M. A. and Rodriguez-Boulan E. (1989) A glycosphingolipid membrane anchor acts as an apical targeting signal in polarized epithelial cells. *J. Cell Biol.* 109, pp. 2145–56.

[165] Paladino, S., Sarnataro, D., Pillich, R., Tivodar, S., Nitsch, L. and Zurzolo, C. (2004) Protein oligomerization modulates raft partitioning and apical sorting of GPI-anchored proteins. *J. Cell Biol.* 167, pp. 699–709.

[166] Paladino, S., Lebreton, S., Tivodar, S., Campana, V., Tempre, R. and Zurzolo, C. (2008) Different GPI-attachment signals affect the oligomerisation of GPI-anchored proteins and their apical sorting. *J. Cell. Sci.* 121, pp. 4001–07.

[167] Scheiffele, P., Roth, M. G. and Simons, K. (1997) Interaction of influenza virus haemagglutinin with sphingolipid–cholesterol membrane domains via its transmembrane domain. *EMBO J.* 16, pp. 5501–8.

[168] Lin, S., Naim, H. Y., Rodriguez, A. C. and Roth, M. G. (1998) Mutations in the middle of the transmembrane domain reverse the polarity of transport of the influenza virus hemagglutinin in MDCK epithelial cells. *J. Cell Biol.* 142, pp. 51–7.

[169] Cheong, K. H., Zacchetti, D., Schneeberger, E. E. and Simons, K. (1999) VIP17/MAL, a lipid raft-associated protein, is involved in apical transport in MDCK cells. *Proc. Natl. Acad. Sci. USA* 96, pp. 6241–8.

[170] Puertollano, R., Martín-Belmonte, F. Millán, J., de Marco, M. C., Albar, J. P., Kremer, L. and Alonso, M. A. (1999) The MAL proteolipid is necessary for normal apical transport and accurate sorting of the influenza virus hemagglutinin in Madin–Darby canine kidney cells. *J. Cell Biol.* 145, pp. 141–51.

[171] Magal, L. G., Yaffe, Y., Shepshelovich, J., Aranda, J. F., de Marco, M. C., Gaus, K., Alonso, M. A. and Hirschberg, K. (2009) Clustering and lateral concentration of raft lipids by the MAL protein. *Mol. Biol. Cell* 20, pp. 3751–62.

[172] Puertollano, R. and Alonso, M. A. (1999). MAL, an integral element of the apical sorting machinery, is an itinerant protein that cycles between the trans-Golgi network and the plasma membrane. *Mol. Biol. Cell* 10, pp. 3435–47.

[173]   de Marco, M. C., Martin-Belmonte, F., Kremer, L., Albar, J. P., Correas, I., Vaerman, J. P., Marazuela, M., Byrne, J. A. and Alonso, M. A. (2002) MAL2, a novel raft protein of the MAL family, is an essential component of the machinery for transcytosis in hepatoma HepG2 cells. *J. Cell Biol.* 159, pp. 37–44.

[174]   Delacour, D., Koch, A. and Jacob, R. (2009) The role of galectins in protein trafficking. *Traffic* 10, pp. 1405–13.

[175]   Delacour, D., Gouyer, V., Zanetta, J. P., Drobecq, H., Leteurtre, E., Grard, G., Moreau-Hannedouche, O., Maes, E., Pons, A., Andre, S., Le Bivic, A., Gabius, H. J., Manninen, A., Simons, K. and Huet, G. (2005) Galectin-4 and sulfatides in apical membrane trafficking in enterocyte-like cells. *J. Cell Biol.* 169, pp. 491–501.

[176]   Mishra, R., Grzybek, M., Niki, T., Hirashima, M. and Simons, K. (2010) Galectin-9 trafficking regulates apical-basal polarity in Madin–Darby canine kidney epithelial cells. *Proc. Natl. Acad. Sci. USA* 107, pp. 17633–8.

[177]   Fajka-Bojaa, R., Blaskó, A., Kovács-Sólyom, F., Szebenia, G. J., Tóth, G. K. and Monostori, É. (2008) Co-localization of galectin-1 with GM1 ganglioside in the course of its clathrin- and raft-dependent endocytosis. *Cell. Mol. Life Sci.* 65, pp. 2586–93.

[178]   Stechly, L., Morelle, W., Dessein, A. F., Andre, S., Grard, G., Trinel, D., Dejonghe, M. J., Leteurtre, E., Drobecq, H., Trugnan, G., Gabius, H. J. and Huet, G. (2009) Galectin-4-regulated delivery of glycoproteins to the brush border membrane of enterocyte-like cells. *Traffic* 10, pp. 438–50.

[179]   Schneider, S., Greb, C., Koch, A., Straube, T., Elli, A., Delacour, D. and Jacob, R. (2010) Trafficking of galectin-3 through endosomal organelles of polarized and nonpolarized cells. *Eur. J. Cell Biol.* 89, pp. 788–98.

[180]   Shenoy-Scaria, A. M., Timson Gauen, L. K., Kwong, J., Shaw, A. S. and Lublin, D. M. (1993) Palmitylation of an amino-terminal cysteine motif of protein tyrosine kinases p56lck and p59fyn mediates interaction with glycosyl-phosphatidylinositol-anchored proteins. *Mol. Cell. Biol.* 13, pp. 6385–92.

[181]   Zhang, W., Trible, R. P. and Samelson, L. E. (1998). LAT palmitoylation: its essential role in membrane microdomain targeting and tyrosine phosphorylation during T cell activation. *Immunity* 9, pp. 239–46.

[182]   Kabouridis, P. S., Magee, A. I. and Ley, S. C. (1997) S-acylation of LCK protein tyrosine kinase is essential for its signalling function in T lymphocytes. *EMBO J.* 16, pp. 4983–98.

[183]   Viola, A., Schroeder, S., Sakakibara, Y. and Lanzavecchia, A. (1999) T lymphocyte costimulation mediated by reorganization of membrane microdomains. *Science* 283, pp. 680–2.

[184]   Varma, R., Campi, G., Yokosuka, T., Saito, T. and Dustin, M. L. (2006) T cell receptor-proximal signals are sustained in peripheral microclusters and terminated in the central supramolecular activation cluster. *Immunity* 25, pp. 117–27.

[185]   Hashimoto-Tane, A., Yokosuka, T., Ishihara, C., Sakuma, M., Kobayashi, W. and Saito, T. (2010) T-cell receptor microclusters critical for T-cell activation are formed independently of lipid raft clustering. *Mol. Cell. Biol.* 30, pp. 3421–9.

[186]   Douglass, A. D. and Vale, R. D. (2005) Single-molecule microscopy reveals plasma membrane microdomains created by protein-protein networks that exclude or trap signaling molecules in T cells. *Cell* 121, pp. 937–50.

[187]   Lin, J., Miller, M. J. and Shaw, A. S. (2005) The c-SMAC: sorting it all out (or in). *J. Cell Biol.* 170, pp. 177–82.

[188]   Davis, S. J. and van der Merwe, P. A. (2006) The kinetic-segregation model: TCR triggering and beyond. *Nat. Immunol.* 7, pp. 803–9.

[189]   James, J. R. and Vale. R. D. (2012) Biophysical mechanism of T-cell receptor triggering in a reconstituted system. *Nature* 487, pp. 64–9.

[190]   Alakoskela, J. M., Koner, A. L., Rudnicka, D., Kohler, K., Howarth, M. and Davis, D. M. (2011) Mechanisms for size-dependent protein segregation at immune synapses assessed with molecular rulers. *Biophys. J.* 100, pp. 2865–74.

[191]   Chen, I. J., Chen, H. L. and Demetriou, M. (2007) Lateral compartmentalization of T cell receptor versus CD45 by galectin-N-glycan binding and microfilaments coordinate basal and activation signaling. *J. Biol. Chem.* 282, pp. 35361–72.

[192]   Kaizuka, Y., Douglass, A. D., Varma, R., Dustin, M. L. and Vale, R. D. (2007) Mechanisms for segregating T cell receptor and adhesion molecules during immunological synapse formation in Jurkat T cells. *Proc. Natl. Acad. Sci. USA* 104, pp. 20296–301.

[193]   Hartman, N. C., Nye, J. A. and Groves, J. T. (2009) Cluster size regulates protein sorting in the immunological synapse. *Proc. Natl. Acad. Sci. USA* 106, pp. 12729–34.

[194]   Bonello, G., Blanchard, N., Montoya, M. C., Aguado, E., Langlet, C., He, H. T., Nunez-Cruz, S., Malissen, M., Sanchez-Madrid, F., Olive, D., Hivroz, C. and Collette, Y. (2004) Dynamic recruitment of the adaptor protein LAT: LAT exists in two distinct intracellular pools and controls its own recruitment. *J. Cell Sci.* 117, pp. 1009–16.

[195]   Griffiths, G. M., Tsun, A. and Stinchcombe, J. C. (2010) The immunological synapse: a focal point for endocytosis and exocytosis. *J. Cell Biol.* 189, pp. 399–406.

[196]   Purbhoo, M. A., Liu, H., Oddos, S., Owen, D. M., Neil, M. A., Pageon, S. V., French, P. M., Rudd, C. E. and Davis, D. M. (2010) Dynamics of subsynaptic vesicles and surface microclusters at the immunological synapse. *Sci. Signal.* 3, p. ra36.

[197]   Cemerski, S., Das, J., Giurisato, E., Markiewicz, M. A., Allen, P. M., Chakraborty, A. K. and Shaw, A. S. (2008) The balance between T cell receptor signaling and degradation at the center of the immunological synapse is determined by antigen quality. *Immunity* 29, pp. 414–22.

[198]   Zech, T., Ejsing, C. S., Gaus, K., de Wet, B., Shevchenko, A., Simons, K. and Harder, T.

(2009) Accumulation of raft lipids in T-cell plasma membrane domains engaged in TCR signaling. *EMBO J.* 28, pp. 466–76.

[199]   Rossy, J., Owen, D. M., Williamson, D. J., Yang, Z. and Gaus, K. (2013) Conformational states of the kinase Lck regulate clustering in early T cell signaling. *Nat. Immunol.* 14, pp. 82–9.

[200]   Itano, M. S., Steinhauer, C., Schmied, J. J., Forthmann, C., Liu, P., Neumann, A. K., Thompson, N. L., Tinnefeld, P. and Jacobson, K. (2012) Super-resolution imaging of C-type lectin and influenza hemagglutinin nanodomains on plasma membranes using blink microscopy. *Biophys. J.* 102, pp. 1534–42.

[201]   Kraft, S. and Kinet, J. P. (2007) New developments in FcεRI regulation, function and inhibition. *Nat. Rev. Immunol.* 7, pp. 365–78.

[202]   Bugajev, V., Bambousková, M., Dráberová, L. and Dráber, P. (2010) What precedes the initial tyrosine phosphorylation of the high affinity IgEreceptor in antigen-activated mast cell? *FEBS Letters* 584, pp. 4949–55.

[203]   Field, K. A., Holowka, D. and Baird, B. (1995) FcεRI-mediated recruitment of p53/56lyn to detergent-resistant membrane domains accompanies cellular signaling. *Proc. Natl. Acad. Sci. USA* 92, pp. 9201–5.

[204]   Wilson, B. S., Pfeiffer, J. R. and Oliver, J. M. (2000) Observing FcεRI signaling from the inside of the mast cell membrane. *J. Cell Biol.* 149, pp. 1131–42.

[205]   Wilson, B. S., Pfeiffer, J. R., Surviladze, Z., Gaudet, E. A. and Oliver, J. M. (2001) High resolution mapping of mast cell membranes reveals primary and secondary domains of FcεRI and LAT. *J. Cell Biol.* 154, pp. 645–58.

[206]   Wilson, B. S., Steinberg, S. L., Liederman, K., Pfeiffer, J. R., Surviladze, Z., Zhang, J., Samelson, L. E., Yang, L. H., Kotula, P. G. and Oliver, J. M. (2004) Markers for detergent-resistant lipid rafts occupy distinct and dynamic domains in native membranes. *Mol. Biol. Cell* 15, pp. 2580–92.

[207]   Wu, M., Holowka, D., Craighead, H. G. and Baird, B. (2004) Visualization of plasma membrane compartmentalization with patterned lipid bilayers. *Proc. Natl. Acad. Sci. USA* 101, pp. 13798–803.

[208]   Ge, M., Gidwani, A., Brown, H. A., Holowka, D., Baird, B. and Freed, J. H. (2003) Ordered and disordered phases coexist in plasma membrane vesicles of RBL-2H3 mast cells. An ESR study. *Biophys. J.* 85, pp. 1278–88.

[209]   Sengupta, P., Holowka, D. and Baird, B. (2007) Fluorescence resonance energy transfer between lipid probes detects nanoscopic heterogeneity in the plasma membrane of live cells. *Biophys. J.* 92, pp. 3564–74.